Introducing Hydrogeology

Other Titles in this Series:

Introducing Astronomy
Introducing Geology ~ A Guide to the World of Rocks (Third Edition)
Introducing Geomorphology
Introducing Metamorphism
Introducing Meteorology ~ A Guide to the Weather
Introducing Mineralogy
Introducing Natural Resources
Introducing Oceanography
Introducing Palaeontology ~ A Guide to Ancient Life
Introducing Sea Level Change
Introducing Sedimentology
Introducing Stratigraphy
Introducing Tectonics, Rock Structures and Mountain Belts
Introducing the Planets and their Moons
Introducing Volcanology ~ A Guide to Hot Rocks

For further details of these and other Dunedin
Earth and Environmental Sciences titles see
www.dunedinacademicpress.co.uk

Introducing
Hydrogeology

Nicholas Robins

EDINBURGH ◆ LONDON

Published by
Dunedin Academic Press Ltd
Head Office:
Hudson House, 8 Albany Street, Edinburgh, EH1 3QB
London Office:
352 Cromwell Tower, Barbican, London, EC2Y 8NB

www.dunedinacademicpress.co.uk

ISBNs
978-1-78046-078-9 (Paperback)
978-1-78046-625-5 (PDF)
978-1-78046-616-3 (ePub)
978-1-78046-617-0 (Amazon Kindle)

British Library Cataloguing in Publication data
A catalogue record for this book is available from the British Library

Typeset by Makar Publishing Production, Edinburgh
Printed in Poland by Hussar Books

Contents

Preface

Hydrogeology is an important and vibrant sub-set of geology. It interacts with a variety of diverse disciplines beyond geology including hydrology, climatology and socio-economics. Hydrogeology is essentially a descriptive science that uses mathematical analysis to provide an essential quantitative component to its output. The mathematics, for the most part, is complex and relies heavily on solving partial differential equations. That being so, this volume describes the basic concepts of groundwater flow analysis in simple language, and avoids burdening the reader with the analytical detail. Nevertheless, all facets of hydrogeology, physical and chemical, are described in the hope that the book will introduce the important role of hydrogeology in underpinning our increasing demands on the environment.

This volume is not intended to be a textbook but rather an introduction to the science of hydrogeology. It also considers a variety of topical issues, for example, climate change impact and water scarcity, nuclear waste repositories and oil-shale fracking. It is hoped that this book will provide insight into a subject that a reader can follow up from more detailed sources, including second degree courses, perhaps even leading to a career as a hydrogeologist.

The text is liberally illustrated with both explanatory diagrams and charts showing real data, as well as numerous photographs. Sourcing the photographic material has relied heavily on the images taken by colleagues and friends, particularly Jeff Davies and Dr Jude Cobbing, who are both former colleagues and good friends, as well as from the Shutterstock image collection. Many of the graphic illustrations are original while others are adapted from published sources. The public domain material available from the United States Geological Survey is also another valuable source of excellent material.

Note: All terms in **bold font** in the main text are defined in a Glossary at the end of the book.

1   Water, Earth's special mineral

What is hydrogeology?

Hydrogeology is the study of underground water in all its facets. It was first defined in the early twentieth century as the study of the laws of the occurrence and movement of subterranean water. Hydrogeology has developed from an early understanding of **groundwater** processes demonstrated by early civilizations in the Middle East, Europe and Asia, and also the Americas. Since the early days of numerical hydrogeology pioneered by the French engineer, Henri D'Arcy (1803–1858) and other contemporary workers, this science has developed into a discipline that can be used to address the complex issues we face today. The term 'geohydrology' is sometimes used. The key is the 'geo' part of the word: an understanding of geology is essential to any investigation of groundwater occurrence.

One significant change in our approach to an understanding of hydrogeology has occurred in the recent past. Whereas hydrogeology was initially a science that could provide solutions to groundwater problems without reference to other disciplines, it has now, quite rightly, become just one component within the larger-scale study of the **water cycle** at sub-catchment, **catchment** and **river basin** scale. Site-specific issues must ultimately be understood in the wider context of the regional basin. Modern hydrogeology therefore provides analysis and understanding of just one component of the water cycle, but it is now pursued within a holistic application that offers solutions that may relate equally to water supply, **ecological or ecosystem services** and the overall management of all available surface and groundwater resources. Indeed, computer **models** can now combine groundwater flow and surface water running at different step intervals. The next challenge for the hydrogeologist will be to make sensible forecasts of water availability within the constraints of current climate change and predictions of demand.

Groundwater

The mineral water is unique on Earth. It is essential to life, it flows in response to gravity and pressure gradients, it is the universal solvent, and it occurs in a variety of phases – for example, as a liquid in oceans, rivers and lakes; as steam from volcanic vents, and as water vapour when it evaporates from open water and from transpiring vegetation. It also occurs in the form of ice in the polar ice caps, which is one factor in controlling the overall seawater level. These phases are essential to the maintenance of the planet. Ice occupies about one-tenth more volume than in its original form as water. This is an important feature because this means that ice is less dense than water and will float upon it. If this were not the case, ice in a lake would sink to the bottom and the entire lake would freeze solid, with a significant cooling effect to the whole planet.

We can see the ice caps, we sail on the sea, we can measure flow in rivers and estimate the amount of water in all the lakes, but one major store of water remains largely unnoticed and largely unseen. This is groundwater, the underground store of water that is contained within the **pore spaces** of sediments, and in cracks and **fractures** in rocks, generally near-surface (Fig. 1.1). This hidden store is half fresh or weakly **mineralized** and **potable** water, and half saline, and it represents 1.7% of all the water on the planet. The fresh component of the groundwater store amounts to an incredible 99% of the global fresh water resource, while lakes, rivers, ice caps and atmospheric water comprise the remaining 1%.

Understanding groundwater, its occurrence, groundwater flow systems and **aqueous chemistry** has evolved alongside our ever-increasing demand for potable water. The development of groundwater understanding over the last 100 or so years has followed three distinct, but related, routes:

1   Elaboration of the relationship between groundwater occurrence, **aquifer systems** and geology, i.e. on the geological setting or underground environment at a range of different scales.
2   Development of mathematical equations to describe the movement of water through rocks and **unconsolidated sediments**.

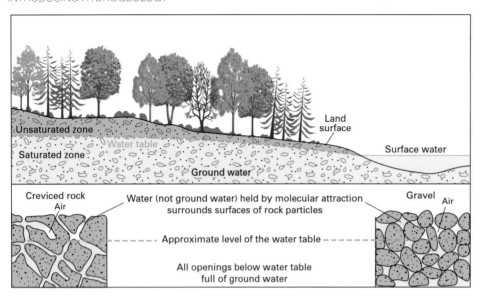

Figure 1.1 The occurrence of groundwater in the saturated zone beneath the water table – note the American spelling of groundwater. Image credit: United States Geological Survey.

3   The study of the chemistry of groundwater, known as **hydrochemistry**.

Our current understanding of groundwater hydraulics has benefited from contemporary developments in data handling and manipulation, which have allowed complex regional and local scale models of groundwater flow systems to be simulated in digital models. These are generally designed for predictive 'what if?' type investigations: what if abstraction from boreholes was doubled; what if the rainfall pattern changed?

The first thread, the geological setting, is vital to understanding groundwater occurrence and groundwater systems. The second, mathematical analysis, enables groundwater flow to be quantified and evaluated numerically. The third allows us to understand the chemical interrelationship between groundwater and the minerals in the containing strata as well as defining chemical types of groundwater, from fresh through brackish to saline.

The hydrological cycle

The hydrological cycle is the continuous circulation of water around the planet (Fig. 1.2). The cycle has neither a beginning nor an end, although the oceans are a useful starting point since they cover the largest area of the planet.

Radiation from the sun evaporates water from the oceans into the atmosphere. The water vapour rises and collects to form clouds. The vapour is almost pure water; sometimes raindrops form around particulate matter, perhaps a chloride-rich particle, becoming acidic. Under certain meteorological conditions the moisture in the clouds condenses and falls back to earth as precipitation, either rain, sleet or snow. This source of fresh water is called **meteoric water**.

The precipitation that falls on the land, rather than back into the sea, is essentially the source of all our fresh water. Some of this precipitation, after wetting any cover of vegetation and the surface of the ground, runs off over the ground surface into ditches, streams and rivers and eventually returns to the oceans. Another part soaks into and penetrates the soil. Much of this water is held in the plant root zone and is returned to the atmosphere by **transpiration** of plants and by **capillary forces** within the soil. Some of it does manage to penetrate below the root zone under the pull of gravity to move downwards until it reaches the water table, where it becomes part of the groundwater reserve.

Capillarity is an important function of water in constrained spaces. It can be seen and demonstrated by holding a small strip of cotton cloth with its lower end dipped into water – the liquid rises up the strip defying gravity. Similarly, water will rise in a narrow tube held vertically with one end just immersed in water.

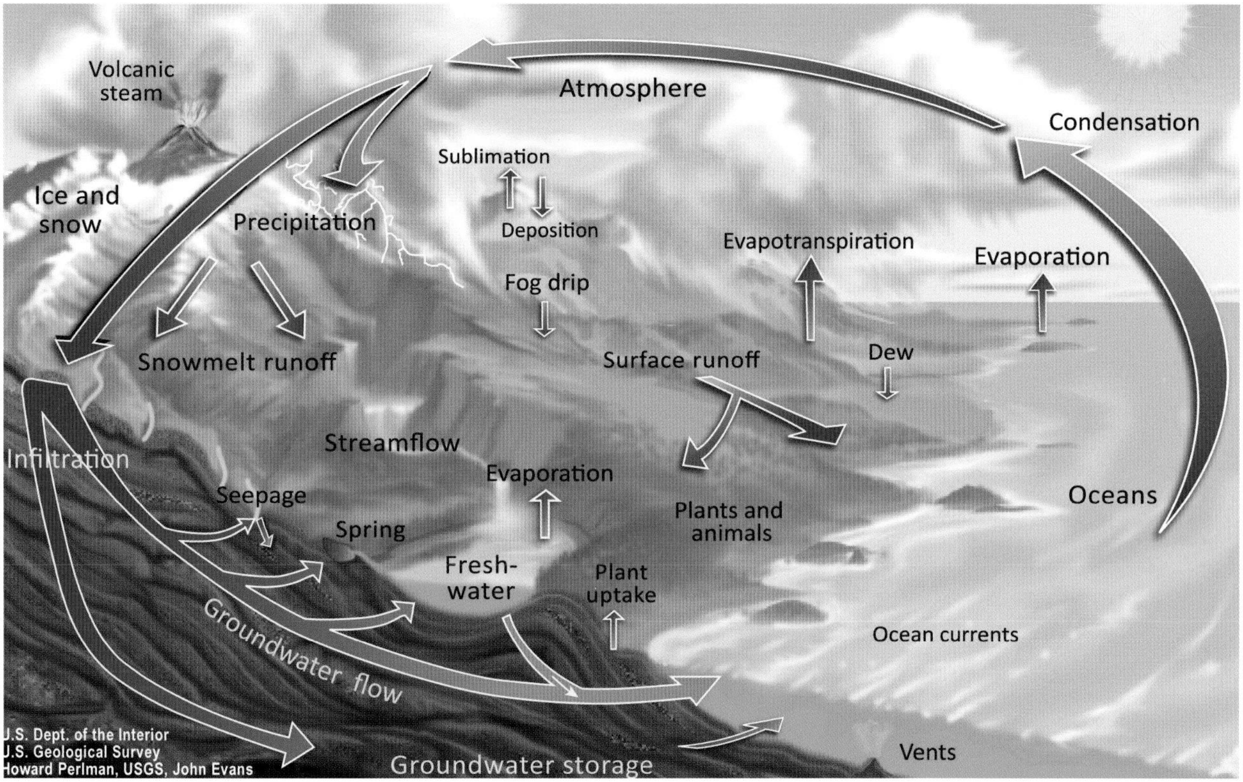

Figure 1.2 Schematic components of the water cycle. Image credit: United States Geological Survey.

Upon joining the body of groundwater, the water that has percolated vertically downwards from the surface can move laterally through the pores of the saturated strata. It may emerge again at the ground surface at lower elevations. Groundwater discharges naturally in the form of **springs, seepages** and **baseflow** directly into surface water bodies such as rivers and lakes, and, of course, at the coast it may flow directly back into the sea (Fig. 1.3). This groundwater baseflow maintains surface flow in streams and rivers during dry periods when surface **runoff** is not available.

The hydrological cycle is thus a continuum that circulates water from the oceans through the atmosphere and allows it to return to the oceans both overland and underground by a variety of pathways. The time taken for a particle of water to complete one circuit of the cycle may vary from just a few hours or days to tens of thousands of years. While the principal driving force is gravity, other forces such as thermal advection,

molecular attraction and capillarity may also influence groundwater flow.

The majority of the moisture in the atmosphere is evaporated from the oceans. Some is also evaporated from terrestrial surface water bodies, such as rivers and lakes, and from transpiration by vegetation. On a cool, still summer evening in places such as Scotland you can sometimes see a mist rising from woodland near valley bottoms – this is the water vapour that derives from the transpiration of the trees. In all about 333 500 cubic km of water are evaporated each year from the oceans and a further 62 500 cubic km per year derive from terrestrial sources.

The cycle is perfectly balanced such that 396 000 cubic km per year is evaporated off the surface of the planet and the same volume falls back on sea and land as precipitation. Of this some 100 000 cubic km of fresh water falls onto land.

4

Figure 1.3 Small spring discharge into a lake in Finland. Image credit: Shutterstock.

Where groundwater occurs

Groundwater occurs in the upper shallow part of the **lithosphere**. The pressure of overburden tends to close fractures and pore spaces with depth so that groundwater may commonly exist down to depths of 100 m but can only circulate below this depth within small dilated cracks associated with major fault systems.

The **water table**, which is not static and may move up and down, is a surface in the ground that divides the unsaturated zone (**zone of aeration**) from the **zone of saturation** (Fig. 1.4). All the pores remain in contact with the atmosphere within the zone of aeration, also known as the **vadose zone**; the zone of saturation or **phreatic zone** is characterized by all the pore spaces, including dilated fractures, being effectively filled with water. The water in this saturated zone is called groundwater.

The unsaturated zone is divided into three parts. The uppermost zone is the soil moisture zone, in which deep roots from trees and other vegetation can draw water back up to the surface. Beneath this is an intermediate zone where the pore

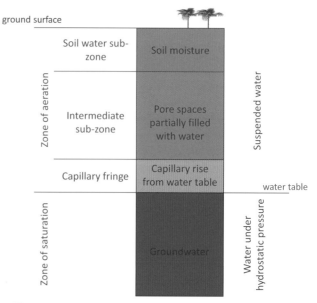

1.4 The zones of aeration and saturation.

spaces are partially filled with water, and at the base of the unsaturated zone is the capillary zone, in which some water can be drawn up from the water table. The capillary zone may be only a few millimetres thick in a coarse sand or gravel, and up to 2 m thick in clay or silt grade material.

In the soil zone both capillarity and molecular forces inhibit gravity from pulling water down from the pore spaces into the intermediate zone. Only when it rains sufficiently for some rainwater to enter the soil profile do the pore spaces contain enough water to overcome these forces and allow water to begin to pass into the intermediate zone towards the water table. This is known as the piston effect. This source of groundwater is called meteoric water and is the most common source of groundwater. There is also **connate water** that was trapped in the rocks at the time of their formation in the geological past, often highly saline water derived from the ancient seas in which sediments were deposited, and **juvenile water** that is drawn towards the surface from great depths in the form of magmatic water.

Condensational water is an important source in arid areas. During the day, land becomes warmer than the air trapped in the soil and creates a pressure difference between the water vapour in the atmosphere and the water vapour in the soil. This draws the atmospheric water vapour into the soil, where it cools and condenses as water. Condensational water is critical for life in desert and semi-desert regions.

Aquifers

A groundwater body contained in rocks beneath the water table is called an **aquifer** (the latin words *aqua* meaning water, and *ferre* meaning to bear). Aquifer is a general term for saturated rock that will yield water in economic quantities, or in arid and semi-arid climates, maintain limited supplies that will sustain small rural communities. In an engineering context all rocks are potential aquifers, as most are able to transmit some groundwater, albeit minute quantities at very slow rates. Aquifers may be small in areal extent and depth: for example, the alluvial valley-fill aquifers characteristic of many upland areas; or they may form large basins such as the Great Artesian Basin in Australia or the Guarani Aquifer that spreads across Argentina, Uruguay, Brazil and Paraguay.

Unconsolidated sand and gravel deposits form valuable aquifer types, which vary in scale from a valley-bottom fluvial deposit to larger alluvial deposits beneath plains, basin

alluvial fanglomerates along mountain fronts, and extensive alluvial deposits and deltas. Valley-bottom aquifers allow water to be drawn both from the aquifer storage and from the adjacent stream or river, which may drain into the aquifer in response to abstraction from **wells** or boreholes (Fig. 1.5). Buried valleys also offer good conditions for **storage** and these are a useful target source in many areas. Sands and gravels beneath large plains may vary in thickness so that their role as an aquifer varies also. Deposits lying at the foot of mountain fronts, such as fanglomerates, benefit from ephemeral stream flows from the mountains to **recharge** the shallow sand and gravel deposits. Occasional wadi flow, for example, off the Montane Plains in northern Yemen and off the ophiolite hills in Oman, illustrate the value of these aquifers in otherwise lowland areas within arid climate zones. Extensive aquifers occur within alluvial deposits underlying lower parts of major rivers and deltas such as the Indus, Ganges/Brahmaputra, Nile and Mississippi.

5

(a) Ground water discharge to a stream

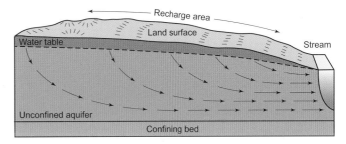

(b) Pumping well intercepting discharge to a stream

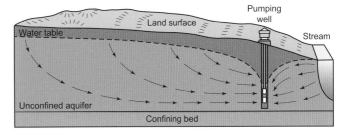

Figure 1.5 (**a**) Groundwater flow to a stream; (**b**) Groundwater flow intercepted by abstraction and locally reversed. Adapted from: *Groundwater Atlas of Colorado*, Colorado Geological Survey Special Publication No. 53.

Groundwater is stored within the intergranular pore spaces of the rock and in any fissures, fractures or other openings, at a pressure equal to the atmosphere at the water table, and at a pressure that increases with depth beneath the water table. This is termed an **unconfined** or **phreatic aquifer**. Groundwater flows laterally from areas of high fluid potential (high **head**), where the water table is highest, to areas of low fluid potential (low head), i.e. it flows along lines of greatest hydraulic gradient, or slope of the water table, from areas where the water table achieves its highest elevation towards those areas where it is at a lower elevation. At the low elevation end of the system the groundwater flows out to surface either as springs and seepages or as valley bottom discharge to streams and rivers or directly to the sea.

Confined aquifers

Rocks that are less able to transmit groundwater may form a hydraulic barrier, or a confining layer to deeper horizons in which water may move more freely (Fig. 1.6). Such is the case in the London Basin, where London Clay confines groundwater in the underlying Tertiary sands and Cretaceous Chalk aquifers, which are termed **confined aquifers**. The sands and the chalk form a confined aquifer beneath parts of London, but are unconfined where they crop out at surface beyond the cover of London Clay towards the higher ground that form the Chiltern Hills to the north and the North Downs to the south.

The pore pressure in the Tertiary sands at the junction with the confining clay is, for the most part, considerably higher than atmospheric pressure. Indeed, most boreholes penetrating the sands were once flowing, artesian wells, with water rising from the borehole to a height dictated by the difference in pressure between the aquifer and the atmosphere; that is, the head in the aquifer was at a greater elevation than the ground surface (Fig. 1.7). However, intensive abstraction throughout much of the nineteenth and twentieth centuries has reduced the head to a sub-artesian state.

In a confined aquifer groundwater flow is caused by differences in pressure between one part of an aquifer and another. If that pressure difference should change – for example, due to a borehole pump being switched on – the instantaneous change causes an elastic response to that change in pressure which is transmitted through the aquifer. This is an important feature of confined aquifers because the pressure change is not entirely due to movement of water, but is mainly due to the elastic transfer of the pressure difference within the aquifer. This may seem

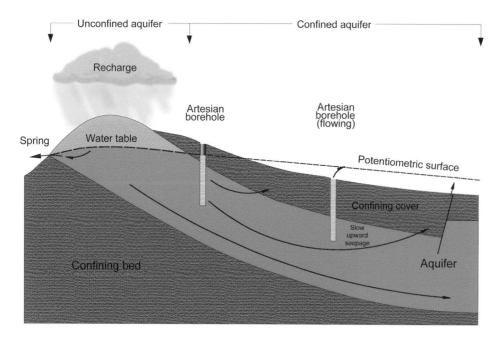

Figure 1.6 Unconfined and confined aquifers showing how artesian flow from a borehole is created. Image credit: UK Groundwater Forum.

Clearly, granular strata that comprise interconnected pores are better able to transmit groundwater than rocks in which the pore spaces are partially infilled with secondary deposits. Thus, an unconsolidated sand with well-rounded grains is more transmissive than an ancient sand body in which calcite or quartz has been redistributed as cement within the pores. It also follows that the larger the pore spaces, the better the strata will transmit water. With small grain sizes of silt and clay grade, capillarity becomes a major factor and water cannot flow easily. Conversely, a karstified limestone, such as that found in the Dinaric Mountains in the Balkans, contains numerous voids (cave systems) large and small, which enable rapid flow of groundwater to drain down the **karstic conduits** from sinks in stream beds, later to emerge at the surface many kilometres away as risings.

Society and groundwater

'Out of sight, out of mind' pretty much describes the attitude many politicians and even water supply practitioners have towards groundwater. An engineer will always prefer an expensive surface water storage and reticulation scheme over a **wellfield**, simply because the resource can be seen and measured directly. But where surface water is not available, at least during the dry season, society has to look to its groundwater resources for supply. The first people to look seriously at groundwater were Asian communities who suffered a scarcity of surface water, lived in large villages and relied heavily on agriculture. These people dug large-diameter wells by hand with primitive tools until they came across the water table. During periods of drought, when the water table dropped, they dug the wells deeper to maintain their supply. They were able to raise the water to ground level by hand or by animal power (Fig. 1.8). Some wells were even large enough in diameter to incorporate spiral stairs for access to the water.

In the arid lands of the Middle East some quite sophisticated engineering was developed to bring groundwater to the surface. The people built long infiltration galleries draining from the higher land to the lower elevations in order to intercept the water table at the higher elevation to drain under gravity down to the village. These are known as kanats, many of them still working, notably in the Gulf states and Iran (Fig. 1.9). The structures could be several kilometres long and drained water both for village use and to support agriculture. They were often sourced from alluvial fans beneath uplands

Figure 1.7 Overflowing borehole in Northern Ireland – the artesian head is marginally above ground level. Image credit: Jude Cobbing.

complicated, but it needs to be remembered because the mathematical descriptions of flow in confined and unconfined states differ significantly.

Fissured aquifers

Groundwater in many aquifers, such as the English Chalk or Permo-Triassic conglomerates and some sandstones, is contained both in intergranular pore spaces and in the many cracks and open joints that characterize these rocks. Although intergranular storage accounts for most of the total volume of groundwater stored in the rocks, the transport potential for groundwater flow is almost entirely dependent on the cracks and joints. These aquifers are called fissured aquifers and sometimes referred to as dual porosity aquifers.

8

Figure 1.8 Camel-powered bucket system in a deep hand-dug well in the Tihama, Yemen.

Figure 1.9 An ancient qanat in Shafiabad village on the Shafiabad caravanserai, Lut Desert, Iran. Image credit: Shutterstock/Pe3k.

as well as soft sedimentary rocks. Kanats were probably first developed in Iran about 2500 years ago, but the technology spread rapidly to Afghanistan and then to the west and to Egypt.

Percussion drilling methods of **boreholes** developed independently both in China and much later in Europe. The Chinese wooden churn drill was powered by hand and was almost identical to similar primitive percussion drills still in use in some parts of the Far East today. Penetration rates were extremely slow, but considerable depths were achieved and many drilled holes reached 50 m depth and beyond. In the twelfth century the same technology was developed in Flanders, where flowing wells had been dug. The artesian flow was even used to power water mills by raising the casing 4 m above ground and cascading the water over a water wheel.

Not until the 1700s was it realized that both surface and groundwater derived from a meteoric source and not from some fantastic distillation of seawater through the ground, while also defying gravity by flowing uphill. The Roman architect Vitruvius had already explained the theory of infiltration of rainwater over the mountains replenishing the lowland groundwater sources, but his theory, correct though it was, did not gain acceptance until much later. The science of hydrogeology has now developed apace, with new understanding being pushed by economic and social demands: the 'Man Made River Project' of North Africa, for example, was a dream come true, although predictions of the longevity of the supply have been grossly over-estimated.

It has been said that development of the hot and arid lands, 'The West' in the United States, was only possible because of three inventions: barbed wire, the borehole turbine pump and air conditioning. True or not, these three things certainly hastened the development of The West with its numerous cattle ranches dependent on groundwater for their survival.

Various groundwater sources have traditionally been credited with medicinal properties, but the only provable example is the purgative effect of water containing Epsom Salts taken into solution from minerals in the surrounding rocks. Clean and potable groundwater is certainly a key to human health, and specific minerals dissolved within it are essential to the human body. However, some minerals such as, for example, fluoride, become toxic if present in excess. The modern fashion for bottled groundwater, often perceived to be associated with health-giving and medicinal properties, for the most part merely offers a safe drinking water.

Today most rural communities in sub-Saharan Africa are entirely dependent on boreholes and hand pumps. Small **yields** of up to 2 litres per second can sustain quite large village communities with clean and potable water (Fig. 1.10). Care is taken to keep animals away from the well-head and

Figure 1.10 This village hand pump in Ghana barely copes with demand; a newly completed borehole is being prepared for the village, but will the discharge be any better? Image credit: Jeff Davies.

to ensure that contaminated surface water cannot enter the borehole. Rural community educational programmes, using puppet shows and graphics, demonstrate the importance of siting boreholes away from pit latrines and safeguarding the likely recharge area of the water source, the so-called **capture zone**, from contaminating activities.

Many large cities are solely supplied by surface water, others by surface and groundwater, while a few are reliant on groundwater alone. Groundwater offers many benefits over surface water:

- It is more commonly free of **pathogenic organisms** and is less likely to need purification for domestic or industrial uses (chlorine may be added, but only to sanitize the water within the pipework of the reticulation system).
- The water temperature is nearly constant – a great advantage for heat exchange users.
- **Turbidity** or suspended fine particulate matter as well as colour are generally absent.
- The chemical character of groundwater is generally constant.
- Groundwater storage is generally greater than surface water storage, providing groundwater with more resilience to short-term drought, although groundwater may take longer to recover post-drought.
- Groundwater is more secure from localized airborne contamination such as radioactive fallout and biological discharge.

Modern society, whether a poor rural community in Africa, a peri-urban community in Asia or a large modern city in South America, is extremely dependent on groundwater. The water supply may be only partly supplied by groundwater, but many of the food stocks consumed are irrigated entirely by groundwater, while the industrial processes that supply our needs may also be dependent on groundwater.

Not surprisingly, there is a close link between **water scarcity**, groundwater, and public unrest. Periodic drought forces people off the land to take refuge in the cities, taxing already stressed urban resources. Mining of groundwater, when abstraction replaces natural long-term replenishment of the resource, places whole cities at risk. Aquifers supplying both Yemen's capital city Sana'a and its major northern regional centre of Taiz are expected to run dry in the foreseeable future; there is no alternative source. Water scarcity was a catalyst of the unrest in Syria, and has long been a critical factor in the tension between Israel and Palestine.

Small island states and communities that live over shared aquifers are also vulnerable. Transboundary aquifers are shared between two or more neighbouring countries and require careful technical investigation and collaborative inter-state management. In countries such as Malawi, where the limited groundwater resource cannot keep up with growing demand, aid organizations continue to drill new boreholes, regardless of an inadequate groundwater resource. The West needs to help all these countries, particularly countries such as Yemen and Syria, where the current instability partly stems from water scarcity issues, and where failure of supply pushes people back into competing tribal factions. Mitigation of the threat to security needs to include technical advice on groundwater management and, where necessary, on preparing for the impacts of projected resource failure.

'Out of sight and out of mind' is an idiom that hydrogeologists have attempted to redress for many decades. Practitioners of hydrogeology play an important role in getting across the message that modern society is very dependent on groundwater and needs to develop this resource to its optimum extent to allow society to achieve its goals. Groundwater resources need to be managed, protected from sources of **pollution** and shared equitably. Groundwater needs to be used in conjunction with surface water in so-called **conjunctive use** schemes, so that groundwater is finally recognized as a foundation of modern society, just as it also underpinned many ancient civilizations.

2 Aquifers and the three Rs: Rainfall, Runoff and Recharge

The water table and the piezometric surface

The water table is the water level surface in an **unconfined aquifer** at which the pressure is atmospheric. It is the level at which the water will stand in a well drilled in an unconfined aquifer. The water table fluctuates whenever there is an imbalance between recharge and outflow from the aquifer. In fact, the water table is constantly in motion, adjusting its surface to achieve a balance between the recharge and the outflow. Generally, the water table forms a subdued version of the topographic features and is deepest below ridges and shallow or reaches the ground surface in valleys. However, sometimes the topographic ridge and the water table ridge may not coincide, and there may be flow beneath a surface divide from one part of an aquifer to another; this is called **watershed** leakage. Wherever the water table intersects the ground surface, a seepage surface or a spring is formed. A perched water table occurs where a small groundwater body is separated from the main aquifer by a relatively small impermeable stratum such as a clay lens (Fig. 2.1). Wells drilled below the perched water table down to the small impervious stratum generally yield small quantities of water and soon go dry.

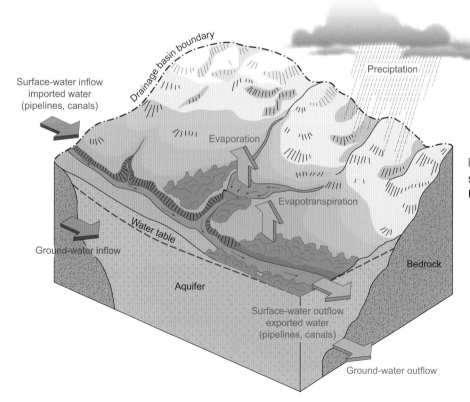

Drainage basin boundary

Precipitation

Surface-water inflow imported water (pipelines, canals)

Evaporation

Evapotranspiration

Water table

Ground-water inflow

Aquifer

Bedrock

Surface-water outflow exported water (pipelines, canals)

Ground-water outflow

Figure 2.1 Conceptual flow model of a small upland catchment. Image credit: United States Geological Survey.

The water in a confined aquifer is under pressure. When a well is drilled in a confined aquifer, the water level in it will rise above the top of the aquifer. This level, the **piezometric surface**, is the level to which the water will rise when a piezometer is inserted in the aquifer at any location, and may be above ground level. It thus indicates the pressure or head of water in the aquifer, and so, for a confined aquifer, the piezometric surface represents the virtual water table.

The water balance and the conceptual flow model

A groundwater basin is a physiographic unit containing one large aquifer, or several smaller connected and hydraulically interrelated aquifers. In a valley between mountain ranges, the drainage basin of the surface stream almost always coincides with the groundwater basin. However, in limestone areas particularly, the surface drainage and groundwater basins may be entirely different, with groundwater flow passing beneath **interfluves** or watersheds from one valley to the next.

Identification of the extent of the groundwater basin is important, as it forms the basic unit for groundwater management. Recharge is expected to balance abstraction and throughflow in each unit in order to maintain groundwater levels without depleting the long-term storage in the aquifer. Continued depletion of the aquifer by over-abstraction is known as **groundwater mining**.

Most aquifers are of large areal extent and may be considered as underground storage reservoirs. In areas such as the United Kingdom, where the geology is complex and outcrop is generally small, the aquifers themselves are also of limited extent. Water enters a reservoir from natural recharge (sometimes also from **artificial** or **engineered recharge**), it flows out under the action of gravity or is pumped out or abstracted in boreholes or wells. In most aquifers the annual throughput of water represents only a small fraction of the overall volume stored in the aquifer. In the Chalk aquifers along the south coast of England the percentage is much greater: as much as 90% of the total store is lost to baseflow discharge or is abstracted in, for example, the Chalk aquifer inland from Brighton.

Each aquifer unit has a **water balance** that describes the long-term equilibrium state of that unit. For each aquifer unit a simple conceptual groundwater flow model can be drawn. The flow model describes the overall groundwater flow pattern for the aquifer unit based on the configuration of the water table, known recharge and discharge zones, and any other information about the aquifer unit available at the start of an investigatory phase.

Groundwater flow conceptualization is the realization of a broad-scale understanding of the hydraulics of any aquifer system. The conceptual groundwater flow model provides not only a preliminary understanding of the groundwater flow system, but also the foundation for a numerical groundwater flow model that can simulate the flow regime under a variety of conditions.

The conceptual model integrates the available knowledge and understanding of the aquifer system. It combines the geological framework, prevailing hydraulic gradients, centres of recharge and discharge and any known boundaries, collectively into a 3D picture of the most likely groundwater flow regime that fits within the known constraints of the aquifer system. This model is usually coupled to an aquifer-wide or catchment-based water balance, in which the effective rainfall over the catchment equates to the combined surface and groundwater outflow inclusive of any groundwater cross-flow, e.g. with other aquifers, and any change in groundwater storage.

The process of developing a conceptual model depends on getting to know the system by identifying its processes, mechanisms and constraints and observing how it responds to rainfall recharge, river stage and abstraction. Data need to be collected, collated and examined, although the understanding at this stage is largely qualitative. The process will be ongoing: some data sets may be incomplete and may not provide appropriate levels of detail. New data can be added at any time, therefore any gaps in the data should also be identified.

The most visible component of the water balance is a spring or seepage. A spring is a concentrated discharge of groundwater to the ground surface. Seepages are smaller volume discharges generally through wet or boggy patches of land or on gently sloping land, for example, located near valley bottoms. In some cases the entire flow is taken up as **evapotranspiration**; in others small streams may flow from a seepage area, while many other seepages are ephemeral. Each spring has its own discrete capture zone, or area in which infiltrating rainwater may eventually discharge back to surface through the spring. There are several categories of spring discharge, all of which are gravity driven:

- Depression springs, where the land surface intersects the water table.

- Contact springs, where a permeable water-bearing formation overlies a less permeable formation at outcrop.
- Artesian springs, where release of water under pressure from confined aquifers discharges where the artesian aquifer outcrops or through cracks in the overlying impermeable strata.
- Fracture springs, where fractures or other voids focus groundwater flow towards the ground surface to a discharge point.

Under favourable conditions, limestone areas can host very large springs. The combination of a high-rainfall catchment and ultra-high transmissive rocks, a rare combination, allows large spring discharges to occur. The ideal stratum is karstic limestone, or dolomitic limestone, in which cracks and fractures, even cave systems, have been created in these brittle rocks over geological time. The enlargement of the fractures is initially caused by solution of percolating rainwater that has been acidified by degeneration of vegetation, which forms organic acids. Rainwater is naturally acid, although poorly buffered, and given suitable cold climatic conditions will enlarge solution features by removing the calcium and bicarbonate in solution while the residual solid detritus forms an abrasive mix to erode or etch along the cracks and fractures. The combination of a high-rainfall spring capture zone and ultra-highly transmissive karstic environments generally occur at or near coastal regions and include, for example, the central Adriatic Coast of the Balkan Peninsular, parts of Vietnam and east central China.

Baseflow

A slightly less visible part of the water balance is seepage or baseflow from the aquifer to streams and rivers where the water table intersects valley bottoms. A third visible component is abstraction points such as boreholes, hand-dug wells and abstraction galleries or horizontal wells pumped via a sump.

Collectively these three features – springs, baseflow to surface waters, and abstraction points – represent the total discharge from the aquifer. Each can be measured directly or indirectly and ratified through analysis of the surface water **baseflow index (BFI)** to identify an overland flow component and the overall baseflow component of the flow derived from groundwater discharge. Baseflow, the groundwater contribution of a surface water **hydrograph** (a graph of water level against time) can also be estimated through a process known as baseflow separation (Fig. 2.2). There are four components to surface water flow:

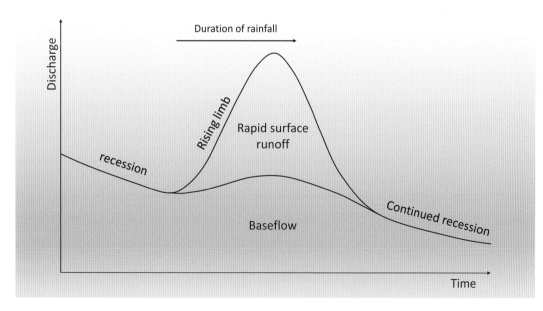

Figure 2.2 Stream flow hydrograph during an intense rainfall event showing likely flow contribution from groundwater baseflow.

- surface runoff,
- **interflow** through the soil to the surface water,
- direct precipitation to the waterway, and
- groundwater baseflow.

Direct precipitation to the waterway ends as the rainfall event ends. For reasons of simplicity, surface runoff is usually combined with soil interflow and direct rainfall onto the waterway. The time base for the runoff phase can be found by drawing a horizontal line through the initial rise of the hydrograph to intersect the recession curve after the rain has stopped. The four components of flow can then be sketched in to reveal the groundwater baseflow component of the surface water flow. There are also complex analytical methods available that provide a better level of confidence for baseflow separation calculations. Note that baseflow directly to the sea in a coastal aquifer cannot be measured and requires more complicated analysis (see Chapter 9).

In some cases the rise in level of the surface water during a rainfall event prevents baseflow from continuing to supply the stream, an **effluent stream**, and the entire flow in the surface water derives from runoff, interflow and direct rainfall when the stream discharges to the aquifer as an **influent stream**, i.e. the elevated stage of the surface water reverses the flow back into the aquifer (Fig. 2.3).

Quite often, the discharge observed from analysis of the surface water flow is equated to the recharge passing into the aquifer to provide a circular water balance. It is obviously better to check this volume by determining the volume of recharge likely to enter the aquifer each year and in the longer term. This can be done by a variety of methods – usually a technique applying soil moisture deficit analysis in humid climates, and one involving **wetting thresholds** in drier climates, where recharge may only occur every so often during cyclonic conditions.

Water balance

Thus, in all humid and maritime climates the three Rs, rainfall, runoff and recharge, dictate a water balance throughout the seasons of the year and also provide useful long-term average values. In arid and semi-arid climates other factors become important, but in either case knowledge of the prevailing climate, land use type and surface gradient can be used to determine recharge to an unconfined or phreatic groundwater body.

Understanding the mechanisms of recharge is the key to groundwater management in order to promote the optimum use of a given resource. Recharge is the component of a rainfall event that reaches the water table and recharges the aquifer

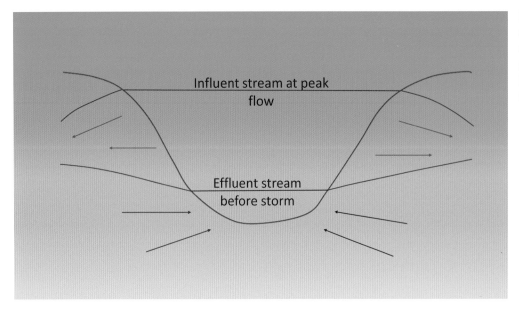

Figure 2.3 Elevated level of a stream changes it from an effluent type, receiving groundwater baseflow, to an influent type with discharge from the stream to the aquifer.

Influent stream at peak flow

Effluent stream before storm

system. In light but persistent rain conditions, recharge may not occur at all, whereas in shorter bursts of heavy rainfall recharge is more likely. How do we know if recharge has taken place? Simply by seeing a rise in the elevation of the water table. If, in the long-term over several years of wet and drier seasons, the water table has persistently declined, then the aquifer is probably being over-exploited. This may be a response to excessive abstraction or to a change in the character of the prevailing rainfall events in response to climate change.

A fine balance is required to maintain the long-term equilibrium of a managed aquifer. The objective is to optimize use of the available renewable resource while maintaining sufficient throughflow to the surface waters to sustain aquatic ecosystems and habitats and to satisfy other riparian demands. This becomes complicated when there are inter-catchment water supplies, and where abstraction causes river water to enter an aquifer, or indeed where leaking sewers and latrines recharge and **pollute** the aquifer.

An aquifer is not defined by water balance alone. Knowledge of the dominant pattern of groundwater flow is also required. This may be input into a preliminary conceptual flow model or entered into a detailed digital groundwater flow model. A conceptual groundwater flow model uses the geometry and overall configuration of the aquifer to create a flow system that is probable and most likely, given the known constraints. This is done as a set of cross-sections or a schematic 3D cartoon. It is an important part of the development of understanding of an aquifer because it puts interpretation of the aquifer on a strictly factual footing.

The conceptual flow model has another important role. It allows us to explore the relationship with other aquifer units and also between topographical watersheds and groundwater divides. For example, an aquifer unit within a large topographical valley is assumed to have a capture zone from recharge that coincides with the topographical divide or watershed. However, the water balance has a surplus of water that cannot be balanced by known discharge and abstraction. The adjacent valley, which is at a slightly higher overall elevation, has a deficit equal to the surplus in the first valley. The conclusion is that the capture zone extends from the first valley beyond the surface-water divide into the second valley. This is critical information for groundwater managers and can be readily obtained from simple analysis of the information provided in

geological and topographical maps without the need for any detailed investigatory fieldwork.

The water balance for an aquifer unit is more complicated when, for example, the aquifer unit drains to a confined aquifer or gains through a weakly permeable floor from a confined aquifer in which the hydrostatic pressure exceeds the elevation of the water table.

The value of the water balance and the conceptual flow model cannot be overstated. These are the foundation to all groundwater investigations, whether aquifer-wide, basin-wide, or on a smaller, local scale.

The application of the water balance and the conceptual flow system is illustrated by a preliminary description of an upland catchment at Gleneagles in Scotland comprising Devonian (Palaeozoic) sandstone, conglomerate and fractured volcanic rocks (Fig. 2.4). Groundwater flows on a catchment scale from beneath the higher ground towards the lower ground to discharge as baseflow into the surface waters. The speed of the flow is a function of the hydraulic gradient, or the inclination of the water table, and the **permeability** or transmissive properties of the rocks (Chapter 3). The volume of groundwater in the flow system depends on the **effective rainfall**, which is the actual rainfall minus **evaporation**, less any surface water runoff.

Superficial strata in the catchment include riverine alluvium along the valley bottom. This is partly clay bound, but also contains some sands and gravels. The catchment is partly overlain by glacial diamicton (mixed clays) at the lower levels.

The sandstone is a mixed bag of weakly permeable strata except where it is locally fractured. In terms of groundwater supply potential, it can sustain borehole yields of one or two litres per second, exceptionally six litres per second. The conglomerate is a preferred target for water supply, with relatively high transmissive properties and storage potential. The volcanic rocks are poorly permeable, although weathering has produced fractures within these strata and intrusive dykes have also encouraged local fracturing.

The water table forms a subdued version of the surface topography, and as such the groundwater divide is directly below the surface water divide. The total capture zone of the upper part of the catchment amounts to some 8 square km. The long-term average rainfall for the catchment is 1265 mm/year. The mean annual loss to evaporation is 416 mm leaving

16

Figure 2.4 The Gleneagles catchment: one of the test boreholes with well-head chemistry kit laid out ready for sampling.

an effective rainfall of 849 mm. The systematic *Hydrometric Analyses*, published by the UK Centre for Ecology and Hydrology, provide a broad overview of the ratio of surface water runoff and groundwater discharge (baseflow) for catchments throughout the United Kingdom; similar data sets are available throughout Europe and North America and in many countries elsewhere. For the catchment upstream of a particular gauging station, the derived baseflow index, the long-term percentage of surface water flow derived from groundwater in the catchment above the gauging station, is given as 0.40 to 0.44, reflecting overland flow over the sandstone and the superficial material. Thus between 40 and 44% of the surface flow derives from groundwater discharge, from both bedrock and the superficial sands and gravels, i.e. about 355 mm equivalent depth over the entire catchment of 8 square km, i.e. about 2.8 million m³ of water.

This water balance for the upper part of the catchment has been derived without any field investigation and can be written in equivalent depth as:

Rainfall (1265 mm) – evaporation (416 mm) – runoff (494 mm) = recharge (355 mm) = throughflow from the upper valley beneath the boreholes (355 mm).

The conceptual flow model (Fig. 2.5) is again derived from basic knowledge of the catchment, the location of the groundwater divides and of the stream, as well as the topography of the catchment. A significant understanding of groundwater quantity and flow direction is now available, without having committed any significant resources. Some understanding of the volume of water available for abstraction has also been gained, although individual borehole yields are likely to be low. This same process can be applied to catchments large and small, and indeed to whole groundwater basins. It is an important exercise and is the key to all groundwater investigation.

Measuring recharge

A practical way of estimating recharge to an unconfined aquifer is the **soil moisture deficit** approach (**SMD**), usually managed as a simple spreadsheet model. **Field capacity** is the amount of water a soil can retain against gravity. Field capacity varies from about 11% water by weight for a sandy loam to 22% for a clay loam. A soil moisture deficit is the depth of rain (mm) needed to bring the soil moisture content of a dry soil back to field capacity when SMD = 0 mm; i.e. the difference

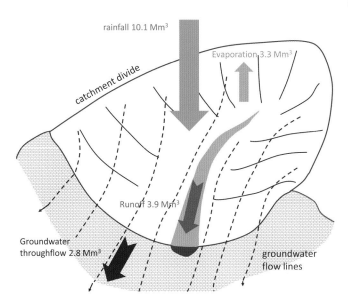

Figure 2.5 Gleneagles conceptual flow model and water budget as volume (million cubic metres), i.e. equivalent depth over the catchment times catchment area of 8 square km.

between the amount of water actually in the soil and the amount of water that the soil can hold. A positive SMD is below field capacity and rain can infiltrate to the capacity of the SMD amount. Saturation is reached when SMD is negative, i.e. there is a water surplus. In a saturated soil all of the available soil pores are full of water, but water will drain downwards out of large pores under the force of gravity.

The soil moisture deficit calculation is based on commonly available meteorological data and land use type (e.g. grassland, crops, trees), and can be run on a daily time step or a larger time interval, depending on availability of meteorological data. The basic calculation is that the difference in soil moisture deficit from one incremental step (e.g. one day) to the next is given by daily precipitation plus daily **actual evapotranspiration** plus the amount draining to groundwater or flowing away overland. Under normal rainfall conditions runoff will be zero, but during intense storm events runoff may be significant.

In many countries the calculation is available from the national meteorological agency based on a 10 km grid or larger. In the United Kingdom the MORECS Data Set is available on a 40 by 40 km grid, while a smaller grid size can also be licensed.

Application of the SMD analysis, for example of the small Scottish catchment, indicates a range in recharge of between 300 and 355 mm. The SMD technique is known to underestimate recharge, particularly in wet and hilly areas, so this range is entirely compatible with the value derived from the water balance estimate above.

In arid climates soil moisture field capacity is rarely achieved and the soil moisture deficit analysis cannot readily be applied. It has been found that runoff is only generated from previously dry soil in an arid or semi-arid area when a specific rainfall threshold is reached, after which the runoff slowly increases up to a maximum 85% of the rainfall intensity. The rate of wetting up is defined by a soil threshold, which is the maximum amount of water absorbed by the soil before any runoff is generated.

A general idea of recharge to an unconfined aquifer can be made with the **chloride mass balance** method. This is one of the most frequently used recharge estimation methods. Chloride is regarded as a suitable environmental tracer since it is highly soluble, conservative and not substantially absorbed by vegetation. The chloride mass-balance method is convenient and inexpensive because of its simple data requirements. The fundamental concept of the method is that the atmospheric input of chloride in precipitation (and a small amount from dry deposition), will be concentrated in residual soil water due to evapotranspiration. Recharge, surface runoff, and evapotranspiration are the paths for water removal from the precipitation area that is considered. Surface runoff is assumed to occur at the ground surface and thus conducts chloride away from the area at approximately the same concentration as is present in precipitation. Although evapotranspiration removes water from the area, it does not remove chloride. Therefore, to preserve the chloride mass balance, chloride concentrations in the recharge water must increase.

Evapotranspiration contains no chloride and if there is no surface runoff, for example in an arid or semi-arid zone, no chloride can escape by that route. The actual processes occurring within the balance zone can be much more complicated than described, but essentially, precipitation volume times chloride concentration equals recharge volume times chloride concentration; i.e. recharge equals precipitation times Cl concentration in precipitation divided by Cl concentration in recharge.

17

Or, expressed as a formula:

[vol. precip. x Cl. conc. = vol. rechar x Cl. conc.

Recharge = precip. x Cl. conc.(precip)/Cl. conc.(rechar)]

Hence, the conventional chloride mass balance approach essentially estimates diffuse recharge through the soil profile. It can be applied to a saturated zone by measuring groundwater chloride or to the unsaturated zone by measuring chloride in soil water. Results are heavily dependent on accurate knowledge of chloride concentrations in rainfall, which may vary between one storm event and another, and on a detailed assessment of chloride distribution in groundwater.

The hydrogeological map and its evolution

The hydrogeological map evolved through the twentieth century into a compact graphic presentation of the hydrogeology of a given area (Fig. 2.6). The central graphic is the map itself, based to a large extent on the available 2D geological map, complete with cross-sections, and accompanied by a variety of inset graphics and text explanations including time series data such as water-level hydrographs and temporal changes in water chemistry. Although UNESCO established a standard international legend in 1970, a range of map types have been produced, including **groundwater vulnerability** and **groundwater potential** maps, which ranged in cartographic quality from simple line work, to complex and highly technical colour printed maps. More recently paper maps have been overtaken by interactive digital maps with a departure away from conventional 2D map format to a user-friendly 3D model format.

In Britain it is convenient to think that the father of the hydrogeological map was Joseph Lucas (1846–1926), but in reality Russian groundwater maps were developing from the

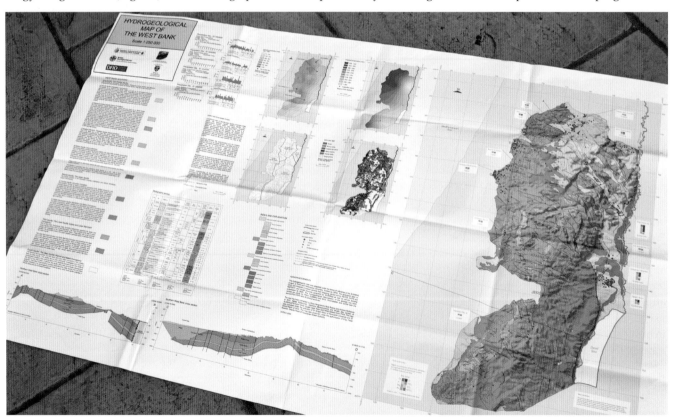

Figure 2.6 The classic UNESCO style Hydrogeological Map of the West Bank illustrating the wealth of information that was presented in these highly crafted documents.

mid-eighteenth century, and both the French and Americans were a little ahead of Lucas. Indeed, work in the Paris Basin is believed to have influenced Lucas in defying his boss, who had charged him with geological mapping in Yorkshire, while Lucas preferred to map groundwater levels in the London Basin!

As early as 2000 BP the people of Asia and Persia were familiar with the concept of hydraulic gradient sufficient to engineer the qanat. Their descriptions of groundwater flow, however, only went as far as vertical schematic slices or cross-sections to illustrate the role of gravity. The first attempt at assembling large-scale groundwater data in map form is credited to the Russian Academy of Sciences, which had been established by Peter the Great in 1724. The Academy directed several expeditions to collect information on natural resources. These were focused on the Ural Mountains, Siberia, the Volga basin, the Kolsky and Kamchatka peninsulas, the Dnieper basin, and the Caucasus. Each expedition collected data that included the occurrence and distribution of fresh water and mineralized springs, groundwater elevations, the relationship between groundwater chemistry and chemical composition of water-bearing formations, and the role of groundwater in karst processes. The results from the first expedition, carried out between 1733 and 1743, were published in 1765 in the monograph *Description of the Land of Kamchatka*, which contained numerous cross-sections and thematic maps with accompanying descriptions of both surface and groundwater features, including mineralized and freshwater springs.

It was a long time before this early form of mapping translated to other countries, and it only did so as need arose. In France maps of the Paris area were published in 1858 showing water level contours in three separate aquifers at a scale of approximately 1: 6700. A larger area of the entire Département de la Seine was mapped to similar criteria shortly afterwards. In North America groundwater mapping was a useful tool for understanding artesian groundwater systems in order to supply essential irrigation schemes without using a steam pump. In Britain, groundwater flow processes needed to be understood when water stored in the Chalk aquifer in the London Basin became a tenable economic asset in the 1870s. Joseph Prestwich (1812–1896) published his seminal cross-sections of the Thames basin 'shewing the extent of the water bearing strata, the position of the more important springs, and the head of water on which those springs and the streams they feed depend'. Joseph Lucas was the first in Britain to gather groundwater level data into contoured piezometry or water level maps. Lucas' map shows, for the first time in Britain, groundwater level contours measured 'early in 1873' and 'at the end of 1873', clearly identifying the transient nature of the water table over an area of about 120 square kilometres.

Time-series data are an important component of the hydrogeological map, although only a few records of weekly or daily manual dipped data commenced in the nineteenth century. The first record of automatic continuous water level recording was that at Oyster Bay in Long Island Sound, New York in June 1903. The equipment comprised barograph stands and a barograph cylinder driven by a double spring clock and a lever, with a pen at one end and a means of attaching a float at the other. It was used to investigate sea tides on a confined coastal aquifer. The length of the lever provided the ratio between the recorded hydrograph and the actual water-level fluctuations.

Groundwater chemistry also developed into a major sub-set within hydrogeology, and both areal distribution of ionic concentrations and variations with time were included in the marginalia of many maps. The maps themselves became ever more complex, some in the mid-twentieth century depicting extent of seawater intrusion, others highlighting the occurrence of mineralized discharges.

A major effort to document resources in Britain began in the late 1940s, and this led to a significant contribution to mapping. By way of example, The Hydro-geological Survey of Kent (sic), carried out by Henry Lapworth Partners for The Advisory Committee on Water Supplies for Kent in 1946, yielded a series of separate thematic maps at a scale of 2 miles to an inch which included depths to the top of the Palaeozoic, Greensand, Gault and Chalk, long-term average rainfall, groundwater level contours, borehole sources and catchment areas.

Another major effort commenced in the late 1970s when the UNESCO International Hydrogeological Map of Europe was compiled and published in collaboration with Bundesanstalt für Geowissenschaften und Rohstoffe: 30 map sheets were produced at a scale of 1: 500 000, each with an accompanying description. Meanwhile the British Geological Survey (BGS) embarked on a rolling programme of hydrogeological map production at scales varying between 1: 625 000 (national scale) and 1: 50 000, the majority of the 23 map sheets being to 1: 100 000 scale.

There are a number of variations on the theme. For example, Groundwater Vulnerability to Pollution maps denote areas were groundwater is at risk and where it is more secure from surface pollutants, based on the likelihood of ingress of the pollutant (Chapter 8). So-called 'Groundwater Harvest Maps' show the likely groundwater development potential at any location within an aquifer or suite of aquifers. Most maps have some accompanying text, usually a summary of the hydrogeological conditions portrayed on the map.

A key problem was how to deal with confined aquifers. The geological map as the basis of a hydrogeological map shows only the rocks at outcrop, some with Quaternary cover, where it is extensive, removed as a second map layer. A confined aquifer, by definition, is not exposed at surface and only features on a geological map in cross-section or textual marginalia. A number of devices were created to deal with this problem: the impressive French 'pyjama stripe'

colouring of a sequence, for example, denoted an unconfined aquifer over a confined aquifer as a single unit. But none of these techniques were wholly satisfactory and all required intelligent interpretation and an understanding of the logic of geometry.

3D interactive maps and models

2D hydrogeological maps fail in general, as do 2D geological maps, simply because they are trying to present a 3D picture in plan form, albeit usually assisted with cross-sections. The 2D geological map, for example, is a mapping geologist's attempt to bring a 3D vision of the pile of rocks being mapped at surface down to a 2D image, which the user then has to interpret to again visualize the whole picture in 3D. For many decades that was the state of technology and little could be done to progress to 3D until computer modelling codes became available in the 1980s (Fig. 2.7). The 2D hydrogeological map was, for the most part, based on the 2D geology map, but with a hydrogeologically meaningful legend

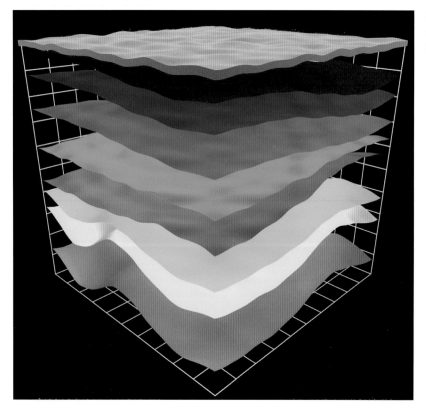

Figure 2.7 3D digital model of an idealized sedimentary sequence. Image credit: © Shutterstock / brumhildich.

in place of the usual lithostratigraphical labels. As such these maps were constrained by the geologist's original portrayal of the 3D system in plan form.

The modern 3D interactive model portraying hydrogeological features within a realistic geological framework is the direct descendant of the conventional geological and general hydrogeological map. Hydrogeological maps have been widely used as the backdrop for **GIS** format databases to facilitate interpretation of data both aerially and according to aquifer type. The power of modern-day **digital imagery** means that there is no longer a need to produce finely crafted and handsomely printed coloured maps. It also means that by linking the digital map to the underlying database, the map or model is automatically updated. These dynamic products can in turn be applied as data platforms to drive a variety of analytical models.

The 3D geological model came first, and then came the issue of how to load hydrogeological data into the geological environment. Planar data, such as piezometry, could be loaded as a surface additional to the geological surfaces within the model, although subject to different constraints. Data on the distribution of physical properties can equally be loaded as 3D planar contours, although sufficient data are rarely available to do this. Much hydrogeological data relate to aggregate values from a borehole column rather than a single point in space, i.e. a borehole as a vertical dimensionless line, the modelled length of which is given by the depth of the borehole; such data are not easy to present graphically. The other problem was how to deal with the fourth dimension so that time-series data could also be loaded, a problem now largely solved through use of the GIS data holding format (see Chapter 4).

3   Aquifer properties and groundwater flow

Porosity

Consolidated rocks vary greatly in **porosity**, water storage and transport properties depending on the degree of consolidation, compaction and cementation, and the development of secondary porosity, such as pressure joints, fractures and solution cracks. The better aquifers are those in which cracks and fractures have developed through solution and attrition into **dilated fracture** systems. Karstic features may develop in which streams and even rivers can disappear down sinks to flow through many kilometres of cave systems to emerge as risings (Fig. 3.1). Solution of the carbonate rock material creates characteristically **hard groundwaters** rich in calcium or magnesium and bicarbonate.

In sediments or sedimentary rocks the porosity depends on grain size, the shape of the grains, the degree of sorting and the degree of cementation. In all types of rock, the porosity also depends on **secondary porosity** or upon the extent, spacing and pattern of cracks and fractures (Fig. 3.2).

The porosity of well-rounded sediments, which have been sorted so that the grains are all about the same size, depends less on particle size than on how tightly the grains are packed.

Figure 3.1 Limestone pavement in North Yorkshire, England, illustrating fracture dilation by solution weathering of the limestone and ultimately of a karst groundwater system.

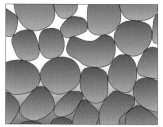

Well-sorted sedimentary material

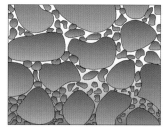
Poorly sorted sedimentary material

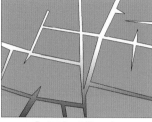
Fractured crystalline rocks

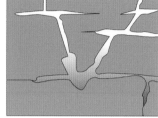
Soluble rock-forming material

Figure 3.2 Extremes of sorting in a granular sedimentary aquifer, in fractured rocks and in dissolved and karst dilated fractured rocks. Adapted from: *Groundwater Atlas of Colorado*, Colorado Geological Survey Special Publication No. 53.

Well-rounded coarse-grained sediments usually have higher porosity than fine-grained sediments. The porosity of sediments is also affected by the shape of the grains. Well-rounded grains may be almost perfect spheres, but many grains are irregular in shape and some are lenticular. They can be shaped like rods, disks, or books. Sphere-shaped grains will pack more tightly and have less porosity than particles of other shapes. The fabric or orientation of the particles, if they are not spherical, also influences porosity. Porosity can range up to 60%. In general, recently deposited sediments have higher porosity.

Dense crystalline rocks or highly compacted soft rocks such as shale have lower porosity, as do highly cemented sedimentary rocks. Poorly sorted sediments (sediments containing a mixture of grain sizes) usually have lower porosity because the fine-grained fragments tend to fill the open spaces. In igneous and metamorphic rocks porosity is usually low because the minerals tend to be intergrown, leaving little free space. Fractured igneous and metamorphic rocks, however, could have high secondary porosity. **Intergranular porosity** (also known as matrix or primary porosity) is the porosity provided by small spaces between adjacent grains

of the rock, and secondary porosity of fractured rocks is the porosity provided by discrete rock mass discontinuities (faults, joints and fractures).

In porous rock, there may be small pores known as dead end pores. These have only one entrance, and so water molecules can diffuse in and out of them, but there can be no hydraulic gradient across them to cause bulk flow of groundwater. In extreme cases, there may be pores containing water that are completely closed so that the water in them is trapped. This may occur during diagenetic transformations of the rock. Porosity usually refers only to pores that can transmit water in the rock; this is termed the **effective porosity**.

A medium that is **homogeneous** has the same packing and grain size throughout; it has the same hydraulic properties throughout its full extent. If there is no difference between vertical porosity and horizontal porosity the aquifer is called **isotropic**. The occurrence of homogeneous and isotropic aquifers is extremely rare – well-rounded beach sand is a possible example. Most aquifer media are heterogeneous and anisotropic and the hydraulic properties vary both horizontally and vertically. Anisotropy occurs when the shape of the grains deviate from a sphere into a flat oval shape. Most sediments are preferentially laid on their flat sides, like tea leaves in a cup, and in the direction of their long axes. Thus, the horizontal permeability or **hydraulic conductivity**, the ability to transmit water in the horizontal plane, is usually greater than the vertical conductivity. Anisotropy can also occur due to facies changes, variations in sorting and deposition of the grains, graded bedding, joints and fractures, clay lenses or variations in cementation.

The amount of water that can drain freely from a porous medium is called **specific yield** and this is always lower than porosity due to capillarity (Table 3.1).

Sandstones and conglomerates are sands and gravels that have been consolidated into hard rocks by cementation of grains, reducing the original pore spaces. This process reduces the intrinsic porosity and hence also the capability to store water. The best sandstone aquifers are those that are fractured or only weakly cemented. Conglomerate aquifers are not common but those that do exist may be high-yielding.

Volcanic rocks also form aquifers; basalts in general offer better conditions for water storage than rhyolites. Permeable zones include erosion surfaces and fossil soil zones between lava flows, flow breccias, shrinkage cracks and joints, as

23

Table 3.1: Typical values of porosity and specific yield for selected media

Medium	Porosity %	Specific yield
Soils	50-60	Variable
Clay	45-55	0.03 or 3%
Silt	40-50	0.05
Medium to coarse-mixed sand	35-40	0.20
Uniform medium-sand	30-40	0.15
Fine to medium mixed sand	30-35	0.10
Sand and gravel	20-35	0.20
Gravel	30-40	0.25
Sandstone	10-20	0.10
Shale	1-10	0.03
Limestone	1-10	0.03

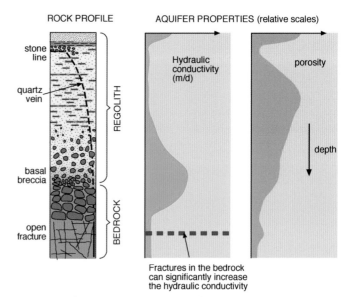

Fractures in the bedrock can significantly increase the hydraulic conductivity

Figure 3.3 Shallow weathering profile for crystalline bedrock into granular regolith and fractured bedrock with indicative permeability and porosity of the weathered strata. Image credit: Jeff Davies.

well as pyroclastic agglomerates. Dolerite dykes may be accompanied by adjacent metamorphic fracture zones in which high-yielding boreholes can be sited to draw on a long, thin, sub-vertical aquifer, generally of limited storage.

Crystalline, igneous and metamorphic rocks offer virtually no interconnected porosity. They can only store and transport water within dilated fractures or within fault zones that contain open joint faces with no fault gouge material. Weathering of these rocks ultimately produces a porous granular **regolith** over fractured bedrock (Fig. 3.3), which in Northern Europe may have been removed by glaciation, though it is a common feature in Africa and Asia.

Clay and silt grade materials, although highly porous, are dominated by low permeability and high capillary retention and yield little if any groundwater even when they are saturated.

Water added to, or discharged from, an aquifer represents a change in the storage volume within the aquifer. In the case of unconfined aquifers this is expressed as the product of the volume of aquifer lying between the water table at the beginning and at the end of a period of time, and the average specific yield of the formation. In the case of a confined aquifer the situation is more complex, as any rise and fall in water level in boreholes penetrating the aquifer reflects a change in pressure rather than in volume. An increase in pressure will induce a small increase in volume, but this is almost negligible. The hydrostatic pressure in the confined aquifer supports the pressure of overburden – the weight of the rocks above it – along with the solid structure of the aquifer

matrix. When hydrostatic pressure is reduced, for example, by pumping water out of the aquifer, compression of the aquifer takes place due to the reduced pressure. This water yielding capacity is termed **storage coefficient**.

The storage coefficient of an aquifer is the volume of water that an aquifer releases from or takes into storage per unit surface area of aquifer per unit change in the compression, or head, normal to that surface. For a square-shaped vertical column of dimensions 1 m by 1 m extending through a confined aquifer, the storage coefficient is given by the volume of water that needs to be released to cause a decline in head of 1 m. In most confined aquifers the value of the storage coefficient lies in the range 0.00005 to 0.005. This reflects the need to induce large pressure changes over an extensive area of the aquifer to produce substantial groundwater yields. Conversely, the storage coefficient of an unconfined aquifer corresponds simply to its specific yield and may be as high as 0.3.

Permeability, transmissivity and storativity

The energy that drives groundwater flow is the force of gravity (Fig. 3.4). Groundwater flows from areas of high head to areas of low head. Groundwater flow depends on the prevailing

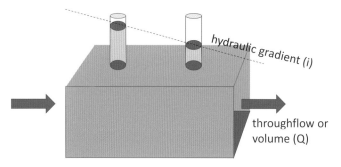

Figure 3.5 Darcy's Law: a unit cross-section of aquifer with hydraulic gradient (i) and throughflow (Q) in which Q is directly proportional to i.

Figure 3.4 Gravity at work: groundwater discharge seepage face from shallow unconsolidated sand aquifer over glacial till, Anglesey, North Wales.

head gradient or hydraulic gradient and ease of flow through the lithologies. This simple relationship is often complicated by factors such as variations in the transmissive properties of the rocks over short horizontal and vertical distances, and by sub-vertical barriers within the aquifer such as clay horizons, volcanic dykes or fault zones, which displace permeable strata against weakly permeable strata. Aquifers are also constrained laterally by weakly or poorly permeable boundaries; faults and the dip of the strata may act as lateral boundaries to the aquifer flow system.

In order to analyse flow in groundwater systems it is necessary to quantify variables, or changes with time, and the hydraulic properties of the aquifer. The two key properties are transmissive capability and ability to store water within interstices such as interconnected pore spaces or dilated fractures. A rigorous and standardized quantification of these parameters allows us to compare the characteristics and performance of different aquifers. It is not essential for the reader to grasp all the basic numerical principles of aquifer flow, but it is useful to try, as these principles are fundamental to a large part of the science of hydrogeology.

The energy required to transmit groundwater through the interconnected pore spaces of an aquifer depends on a function called hydraulic conductivity or permeability (k). This has the length (L) and time (T) dimensions $L^3 T$, L^{-2} and relates rate of groundwater flow Q^{-1} per unit area of aquifer, A,

with the prevailing hydraulic gradient, i, according to **Darcy's Law**, whereby Q is given by the product kiA. (Fig. 3.5)

Permeability
Hydraulic conductivity, or permeability, depends not only on pore size and interconnectivity of the pores, but also on the **viscosity** of the groundwater. With the exception of deeply circulating hot brines, the viscosity of groundwater is normally assumed to be constant. However, the properties of the aquifer are sometimes quoted as **intrinsic permeability, k'**, which relates solely to the rock properties and has the dimension L^2. The relationship between the two parameters is given by $k' = kv/g$, where v is the viscosity of the groundwater and g is gravitational acceleration.

Transmissivity
Transmissivity, T, is the product of hydraulic conductivity (k) and effective aquifer thickness (b) for a single homogeneous formation, or the integral of the product in a layered aquifer. This allows comparisons to be made between aquifers or saturated horizons and their potential to transmit water.

Storativity
The ratio of the volume of water released under free drainage to the bulk volume of the rock from an unconfined aquifer is termed the specific yield, Sy. It has no units, as it is a ratio and values are generally less than 10^{-1} (Table 3.2). In a confined aquifer, where water is released as a result of an elastic response, the ratio of water to rock is called **storativity, S**. Clearly, less water will be released from a confined aquifer,

Table 3.2: Typical aquifer properties

Rock type	Hydraulic conductivity (ms^{-1})	Specific capacity	Specific yield	Comments
Gravel	$10^{-2} - 10^{-1}$	0.20 – 0.30	0.10 – 0.25	
Sand	$10^{-4} - 10^{-3}$	0.25 – 0.40	0.10 – 0.25	
Silt	$<10^{-5}$	0.35 – 0.45	0.05 – 0.10	
Consolidated sand/silt	$10^{-5} - 10^{-4}$	0.20 – 0.40	<0.10	
Sandstone	$<10^{-4}$	<0.25	<0.20	Partly fissure flow
Limestone	$10^{-5} - 10^{-3}$	0.05 – 0.30	<0.10	Mainly fissure flow
Karst limestone	$10^{-2} - 10^{-1}$	<0.35	<0.15	Channel flow
Igneous extrusive	$>10^{-4}$	<0.10	0.05 – 0.10	Fissure flow
Igneous intrusive	$<10^{-4}$	<0.02	<0.01	Fissure flow

and values of storativity are correspondingly lower, usually less than 10^{-3}.

There is also such a thing as a **dual porosity** aquifer. This is an aquifer that has both an intergranular matrix in which the majority of the water is stored, and an interconnected system of dilated cracks and joints in which the groundwater flow takes place. The chalk aquifer of England is a dual porosity aquifer, in which water drains or seeps from the rock matrix into the cracks and joints as head decreases during the drier summer season, and drains back into the matrix as head increases when recharge occurs from late autumn onwards. The cracks and joints are fundamental to the flow system, and understanding of the drainage process caused by the properties of both the porous rock matrix and the secondary porosity of the cracks and joints is crucial to assessing such aquifers.

For those readers who do not want to get too involved with the mathematical side of groundwater understanding, aquifer properties can also be estimated by following a number of 'rules of thumb'. Although their mathematical justification may not be exact, they are valid, for the most part, given certain constraints. For example, an informed guess of the properties of an aquifer can be made based on previous experience with similar rock types. It is also possible to relate the physical properties of an aquifer to probable values of intergranular permeability: soil density obtained from a standard penetration test may be used to obtain estimates of intergranular permeability in terms of granular and cohesive soils and dense, medium-dense and loose soils. Hydraulic conductivity can also be estimated using an empirical relationship with the effective grain size (d_{10}), provided the sediment is well sorted.

Gravity and groundwater flow

Groundwater flow beneath the water table is almost entirely driven by the force of gravity. An element of **thermally induced flow** may also occur in deep aquifers, but otherwise groundwater flows from recharge areas of high head to discharge areas of low head. In a confined aquifer the same principle applies, and groundwater will flow in the direction of declining pressure, again usually from a recharge area of an unconfined aquifer towards a discharge area, which may comprise an abstraction wellfield or may even be a natural discharge zone through a leaky confining layer.

Groundwater flow-paths follow a three-dimensional course, although they are often shown diagrammatically in two-dimensional plans or cross-sections (Fig. 3.6). Flow in a large basin tends to be arcuate such that the longer flow-paths follow the deepest circulation while the shorter flow-paths take a shallower line of circulation. The **circulation time**, the time for a particle of water to travel the length of a flow-path from start to finish, dictates the age of the water at its discharge point. Converging flow-paths to points of emergence such as springs and boreholes provide a cocktail of different ages of water so that precise dating of groundwater at the point of discharge may be difficult (Chapter 7).

Groundwater flow can be analysed using a **flow-net**. Because no flow crosses an impermeable boundary, flow-paths must lie parallel to it. Similarly, because the water table forms an upper boundary of groundwater flow, it too lies parallel to the nearest flow-path except when recharge occurs to the water table. Mapping groundwater flow in two dimensions becomes quite straightforward, given knowledge of the topography of the water table. The water table generally follows a

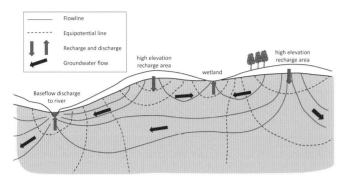

Figure 3.6 Shallow groundwater flow paths and equipotential lines.

subdued version of the ground surface topography. This is caused by the combined effect of infiltration from rainfall over the higher ground and groundwater drainage and discharge to streams at lower elevations. In a confined aquifer, the highest elevation obtained by the potentiometric (piezometric) surface must lie beneath the elevation of the recharge area where the aquifer is at outcrop and locally unconfined. The lowest point of the potentiometric surface is given by the lowest discharge point from the confined aquifer.

In general, hydraulic conductivity and storativity are greater, and depths to the water table are less, beneath valleys than beneath hills. The main reasons for this are:

- Valleys tend to be eroded into rock that is more permeable than elsewhere because of the occurrence of dilated joints and other discontinuities produced through chemical dissolution.
- Stress release as a result of the removal of overburden in the valleys may enhance permeability through the development of cracks and subsequent weathering.
- Groundwater flows from highland towards lowland areas, but a groundwater divide need not necessarily coincide with a topographical divide. Lowland areas can thus draw on a greater volume of storage derived from the total catchment area than can sites at higher elevations in the same catchment.

On a site-specific basis, a simple walk-over survey and a limited programme of exploratory drilling may often yield sufficient information to find the nature and direction of groundwater flow. Three shallow boreholes, each located near the corner of an equilateral triangle, reveal

the direction and gradient of the water table, and brief pumping tests (Chapter 5) can provide values for hydraulic conductivity and storativity, which in turn give an approximate rate of flow. However, on a larger scale several other important factors need to be considered. These include the identification of local and intermediate flow patterns, their recharge and discharge zones, and the calculation of the total throughput of groundwater. As spring discharges indicate the elevation of the water table wherever they occur, the position of the water table along valley sides is revealed by the spring lines.

More often than not, the water table and the surface waters in valley bottoms are at one and the same head, with groundwater discharging as baseflow, or in some cases gaining from the surface water. Thus, the elevation of streams and rivers over an unconfined aquifer coincides with the elevation of the lowest section of the water table. Spring elevations and depth to water in boreholes and wells that are not being pumped provide additional information about the shape of the water table; geophysics may provide even more.

Once the water table elevation data have been compiled, they can be contoured to provide equipotential lines (see Fig. 3.6). Because they are equipotential, all flow paths must be perpendicular to these lines. Furthermore, if the volume of groundwater flow between each schematic pair of flow lines is constant, as given by area times velocity, then a flow-net can be readily constructed, as shown. Note that flow cannot cross the flow-lines. The flow-net provides valuable information regarding the more favourable locations to site new boreholes, i.e. where the permeability is greatest. Wherever the equipotential lines and the flow-paths are closest together, then the permeability is inevitably higher than where they are more distant. In general, the permeability is proportional to the inverse hydraulic gradient or the area contained between adjacent flow-lines and equipotential contours.

Real three-dimensional problems are not so easy to evaluate and solve. However, useful solutions can be obtained by considering a number of vertical cross-sectional segments. Analysis can also be carried out by numerical simulation using the finite difference method or the finite element method (Chapter 4).

Circulation and residence times

Large basins

Groundwater flow systems can be local, intermediate or regional in extent. Local flow occurs between adjacent recharge and discharge zones in areas of pronounced relief (e.g. most shallow unconfined flow in the UK is on a local scale). Intermediate systems are more extensive and encompass several local systems. Regional flow systems, however, extend across larger groundwater basins from the principal watershed to the lowest discharge point. Regional systems cannot develop within the geographical constraints of the UK, but they do occur within most continental sedimentary basins. A regional flow system may contain several nested intermediate systems, and these intermediate systems may contain nested local systems.

A typical regional system of continental scale is the Great Artesian Basin of Australia. This is a confined groundwater basin in which a complex, layered aquifer system is developed within sandstones, siltstones and mudstones of Triassic to Cretaceous age. Precipitation falling on the Great Dividing Range rapidly drains eastwards to discharge at the coast along flow-paths 100 to 200 km in length. The surface water flows west from the Great Dividing Range along intermittent water courses, then infiltrates the stream beds to join the major regional groundwater flow system that spans the Great Artesian Basin. Discharge from the regional groundwater flow system occurs either to the inland drainage of Lake Eyre or to the Darling River (Fig. 3.7). Individual groundwater flow-paths are up to 1000 km in length. **Radiometric dating** (Chapter 7) shows that the age of the groundwater increases down the

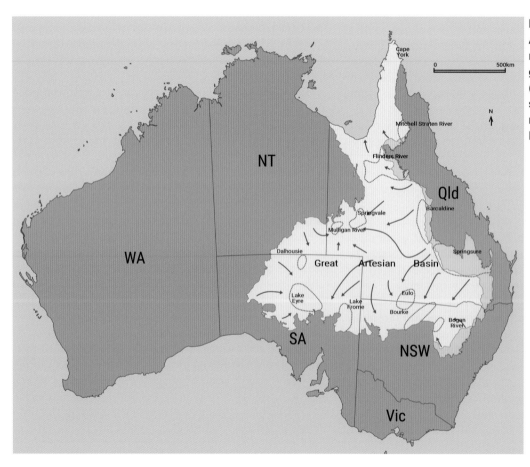

Figure 3.7 The Great Artesian Basin showing recharge areas (yellow), general flow direction (green arrows) and main spring areas (bounded by red lines). Image credit: Friends of Mound Springs.

flow-paths to a maximum age in the vicinity of Lake Eyre of some 2 million years BP (before present).

The Ogallala Aquifer is another very large regional aquifer within a continental scale basin. This aquifer is a largely shallow, unconfined aquifer system in sand, silt, clay and gravel beneath the Great Plains of the United States. The Ogallala Aquifer varies in depth, and is deepest where it fills palaeo-valleys and channels. It is one of the largest aquifers on Earth, and it underlies an area of approximately $450\,000\,km^2$ within parts of South Dakota, Nebraska, Wyoming, Colorado, Kansas, Oklahoma, New Mexico, and Texas. The aquifer is a component of the High Plains Aquifer System, and overlies the Ogallala Formation, which is the principal geological unit underlying 80% of the High Plains.

The Ogallala Formation consists mostly of coarse sedimentary rocks in its deeper sections, grading upward into finer-grained material. The saturated thickness ranges from a few metres to more than 300 m and it is deepest in the Northern Plains. The depth of the water below the surface of the land ranges from almost 120 m in parts of the north to between 30 and 60 m throughout much of the south. Present-day recharge of the aquifer is small and cannot keep pace with abstraction. This indicates that much of the groundwater is **palaeowater**, and radiometric dating recognizes a component of the water that dates from before the last Ice Age, mixed with a smaller volume of modern recharge. Groundwater within the Ogallala generally flows from west to east at an average rate of a 0.4 m/day. Hydraulic conductivity lies in the range 7.5 to 91 m/day.

Large-scale extraction for agricultural purposes only started in the early 1950s. Today about 27% of the irrigated land in the United States lies over the Ogallala aquifer, which yields about one-third of the groundwater used for irrigation in that country. The aquifer, however, is at risk from over-abstraction and pollution by agrochemicals (Fig. 3.8). Agricultural irrigation has reduced the saturated volume of the aquifer by about 10%, reducing water levels by up to 3 m. Once

Figure 3.8 Groundwater sprayed from a centre pivot irrigation beam in western Kansas. Image credit: Shutterstock/ Eugene R Thieszen.

depleted, the aquifer will take over 6000 years to replenish naturally through rainfall, and agricultural activity now needs to downscale. The aquifer system also supplies drinking water to most of the 2.5 million people who live in the High Plains.

Smaller basins – the UK

In contrast to the large continental groundwater basins, the maximum age of shallow groundwater circulating in local or intermediate flow systems in the UK ranges from only a few hours to a few tens of years. Exceptionally, water issuing from the thermal springs at Bath, southwest England, which has a discharge temperature of 46.5°C, penetrates to a depth of about 2 km from its recharge area in the Mendip Hills, and takes nearly 10 000 years to complete the journey from the Mendips to Bath. Shallower flow paths, for example in the chalk aquifer within the London Basin, have yielded radiocarbon dates of up to 25 000 years in selected areas. Elsewhere, groundwater ages in the UK are typically measured in tens, or at most hundreds of years. However, the brines trapped in the eastern part of the Lincolnshire Limestone aquifer represent fossil or connate water – concentrated seawater dating from when the sediments were laid down at the bottom of the Jurassic sea.

Changed flow patterns

Intervention with the natural flow in an aquifer can induce groundwater to move in a new direction. For example, this can be caused by pumping from a well or borehole or conversely, and sometimes unintentionally, by artificially recharging an area of aquifer by means of recharge lagoons or injection boreholes (Chapter 6). The induced flow patterns caused by localized changes in head are overprinted on the naturally existing flow pattern, and may have the effect of diverting groundwater from one area to another. Induced flow systems may be capable of moving large volumes of groundwater, but they are generally only effective over small areas of an aquifer system.

Groundwater interactions

Surface water is an integral part of all groundwater flow systems. Groundwater interacts with small upland streams, lakes, and wetlands in headwater areas, with major river valleys and with the coast. It is generally assumed that higher elevation areas are the recharge zones for groundwater, and that lower elevation areas are groundwater discharge zones, but this is really only the case in regional flow systems. The superposition of local flow systems associated with surface-water bodies on this regional framework results in complex interactions between groundwater and surface water in all landscapes, regardless of regional topographic position. Hydrologic processes associated with the surface waters, such as seasonal variation in water levels, along with evaporation and transpiration of groundwater from around the perimeter of surface-water bodies, influence seasonal changes in the groundwater flow fields near to a surface-water body. Thus, a groundwater system may discharge into the surface water in the dry season but may gain from the river in the wet season when river stage is at its highest.

Non-basinal aquifers

There are a few aquifers of very large extent that are not basinal. The Precambrian crystalline basement rocks that underlie much of southern Africa within igneous cratonic and metamorphic mobile belt/orogenic provinces are non-basinal. In the unweathered state, these strata are regarded as non-aquifers. However, limited quantities of groundwater found in weathered and fractured zones are used to sustain water supply to rural communities. Although basement groundwater resources are widely developed using springs, wells and boreholes, their occurrence in weathered and fractured horizons is complex.

Semi-arid areas

Shallow weathered regolith and the underlying fractured crystalline basement bedrock form the main source of water for many rural communities in otherwise drought-prone sub-tropical, savannah and semi-arid areas (Fig. 3.9). Typical well yields are small, usually less than 2 litres per second, but are adequate to sustain rural communities from a hand-operated pump. Despite the areal extent of the aquifer, there is little if any regional groundwater flow. The aquifer is shallow and contained within the weathered regolith of the crystalline bedrock, with local recharge and discharge zones.

Rainfall within this semi-arid region of Southern Africa is seasonal and episodic with distinct wet and dry cycles. Climate variability now tends to perturb this regularity, although clear wet and dry periodic cycles remain. Groundwater recharge mainly takes place during the wet cycle, a period of five or

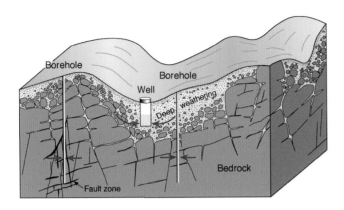

Figure 3.9 Basement aquifer with local flow systems and small source capture zones. Regional flow systems tend to be absent or of low volume. Image credit: Jeff Davies.

more years, when rainfall sustains crop growth, animal husbandry, and community wellbeing. During the dry climate cycle, surface waters diminish. After the first year of the dry period, families and their crops and livestock become solely dependent on groundwater drawn by village hand pumps and scarce motorized water-points. The sustainability of the supply, until the onset of the next wet climate cycle, depends on the groundwater available in storage in the shallow aquifer at the start of the dry climate cycle. Its failure heralds the onset of severe water scarcity and real hardship for rural communities.

Groundwater flow and numerical analysis

The mathematical theory of groundwater flow

The basic **groundwater flow equation** is the mathematical relationship describing the flow of groundwater through an aquifer. **Transient flow** of groundwater is described by a form of the **diffusion equation** (similar to that used in heat transfer, describing the flow of heat in a solid by conduction). **Steady-state flow** of groundwater is described by a form of the Laplace equation, which is a form of potential flow and has analogues in several other fields. Thus, there are two basic equations that govern the flow of groundwater through a porous medium: Darcy's Law, which we have already come across in Chapter 3, and the conservation of mass.

Darcy's Law

Darcy's Law can be written in a variety of forms that can be applied to different situations. Essentially it relates the velocity of flow to the permeability of the medium and the prevailing head difference across a porous medium. In its simple form it is given as:

$$Q = -kA \, dh/dl$$

Where Q is flow volume across a cross-sectional area A subject to a pressure difference (hydraulic gradient) of dh/dl in a porous medium with a permeability k.

Darcy's law is a simple mathematical statement that neatly summarizes several familiar properties exhibited by groundwater flowing in aquifers, including:

- if there is no pressure gradient over a distance, no flow occurs (these are hydrostatic conditions),
- if there is a pressure gradient, flow will occur from high pressure towards low pressure (opposite to the direction of increasing gradient – hence the negative sign in Darcy's law),
- the greater the pressure gradient (through the same formation material), the greater the discharge rate, and
- the discharge rate of fluid will often be different – through different formation materials (or even through the same material, in a different direction) – even if the same pressure gradient exists in both cases.

Conservation of mass

The **mass balance equation** is simply a statement of accounting: that for a given control volume, aside from sources or sinks, mass cannot be created or destroyed. The conservation of mass states that, for a given increment of time (Δt), the difference between the mass flowing in across the boundaries, the mass flowing out across the boundaries, and the sources within the volume such as recharge, is the change in storage:

$$\Delta M_{sto}/\Delta t = M_{in}/\Delta t - M_{out}/\Delta t - M_{rech}/\Delta t$$

If the aquifer has recharging boundary conditions a steady-state may be reached (or it may be used as an approximation in many cases), and the diffusion equation (above) simplifies to the Laplace equation.

The mass balance equation has also been seen before in the description of flow-nets. For each flow tube contained between two adjacent flow-lines, the volume of water coming in is equal to the volume of water coming out. Indeed, a common way of deriving an estimate of the Laplace equation is by drawing a flow-net, or by flow-net analysis, as it is called. The answer will not be precisely accurate but it will provide a useful working volume.

Most analytical formulae used in hydrogeology can be derived from Darcy's Law and the mass balance equation. It is not necessary to fully understand the derivations because most can be applied in spreadsheet models that provide a usable output (Chapter 5). However, in order to model groundwater flow in an aquifer or a part of an aquifer, we need to understand the variables that apply to that aquifer. Apart from the **formation constants**, permeability and storativity, a history of change in storage is essential to calibrate the model.

Change in storage – the hydrograph

Changes in groundwater storage in an unconfined aquifer require a corresponding change to the elevation of the water table. Changes in storage in a confined aquifer require a corresponding increase or decrease in hydrostatic pressure. Therefore the water level in a borehole penetrating below the water table in an unconfined aquifer, or in a borehole

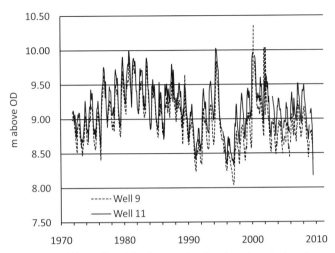

Figure 4.1 Water level hydrographs for two boreholes in an unconsolidated aquifer at Ainsdale, Lancashire, 1972–2010, showing cyclic trends with clear wetter and drier periods.

Figure 4.2 An electric dipper being used to measure depth to water while pumping at low yield from a newly completed borehole in Scotland. Image credit: Jude Cobbing.

penetrating through the overlying confining layer of a confined aquifer, will rise and fall due to change in storage (Fig. 4.1). In the winter, or in the wet season, the water level will rise due to recharge entering the unconfined aquifer and increasing the head on any related confined system. In summer when rain is scarce, the water level will recede.

Taking measurements

Spot measurements of the depth to water level can be made with an electric dipper. This is an electrode suspended on a graduated tape. When the electrode touches the water surface the dipper reel emits a beeping sound. The depth to the electrode and the depth to water can be read off the graduated tape (Fig. 4.2). Note the datum used for consistency; it might be the top of the borehole casing, it might be some other fixed point. Time series data can be recorded using a pressure transducer suspended down a borehole to some point beneath the water level. There is also a data logger in the transducer capsule, and this can be set to record the head of water above it at 10 second intervals or at longer intervals up to hourly. Once raised and downloaded onto a portable computer, several months later, the data will plot automatically as a time series hydrograph, as depth to water against time. It is important to measure the depth to water each time the transducer is set, to ensure that real depth data are recorded, not just variations

in head. The data can also be plotted against sea level if the elevation above sea level of the datum (usually top of the borehole casing) is known.

Plotting data

Borehole hydrograph data can be plotted against rainfall data, usually as daily rainfall rather than actual rainfall events. This allows a graphic presentation of the relationship between rainfall and recharge, lack of rainfall and recession, and provides a useful constraint on recharge estimates derived by other methods. The first example shown (Fig. 4.3) illustrates the relationship between depth to water table in alluvial gravels in the Thames valley, rainfall, and ephemeral spring discharge when the water table rises above the ground level. The gravel deposits are fed from the Chalk hills to the east side and discharge is constrained by the River Thames once it floods overbank to the west. The second example (Fig. 4.4) is from a groundwater hydrograph in South Africa following the 1992 drought. Post-drought rainfall is struggling to replenish the aquifer year by year. In this case replenishment did not occur for a further 15 years, during which time the groundwater resource was depleted and incapable of satisfying demand.

Some confined aquifers are so tightly confined that they act like a drum. Some will respond to pressure changes due to Earth tides, some even to the pressure waves from distant

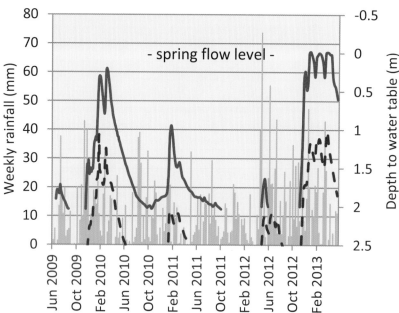

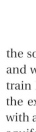

Figure 4.3 Borehole hydrograph and rainfall in the Thames Valley fluvial deposits; note occurrence of ephemeral spring flow as the water table reaches a set level.

Figure 4.4 The 1992 drought in South Africa. Water levels did not recover the pre-drought levels for 15 years. Image credit: Jeff Davies.

earthquakes. Perhaps the oddest example is a box-shaped rise and fall in a borehole recorded over a small confined aquifer in Botswana. The water level consistently went up by 3 cm every morning at quarter past two and went down to its former level at about half past three. Nobody understood this until a railway enthusiast looked at the data. The main single track railway from Mafeking to Bulawayo straddles the aquifer and there is a siding for trains to pass. Every morning the southbound goods entered the siding at quarter past two and waited for the Bulawayo Express to rush past. The goods train left at half past three to trundle south again, removing the extra mass of the train from the confining surface layer, with a corresponding reduction in pressure of the water in the aquifer!

Predictive modelling

Simulation of groundwater flow systems has progressed from the membrane models and electrical resistance models of the 1960s, through the early computer programs that ran on mainframe computers, to modern-day computer code that can be run on a lap-top. So automated has the modelling process become that there is a real fear that human interaction with the output may become overlooked.

A groundwater model is only as good as the data put into it. Some idea of the reliability and representativeness of the data is needed in order to appreciate the uncertainty of the output. This requires some understanding of the aquifer itself; it certainly requires a walk-over visit guided by a hydrogeologist who has knowledge of the area and who can explain that the borehole data may not be accurate, or the recharge estimate may be in error. By way of illustration, a well-known, newly inaugurated professor of hydrogeology at a top flight British university was taken into the field to a drilling site by one of his students. A mathematician and groundwater modeller of some standing, the Prof declared, 'So that's what a drilling rig looks like!'.

Limitations of computer models

The other problem with groundwater modelling is that no matter how inaccurate the input data are, the output end is slick, colourful, looks highly professional and is believable. But a model without an uncertainty label is to be treated with caution. Throughput calculations should always be checked by a simple Darcy calculation to ascertain the order of the flow volume. So often a model produces a number that is just not possible. The other problem is the modeller's passion for changing the data. The rainfall data doesn't fit: let's reduce rainfall by 20% even though we know it is correct and has been measured directly. Surely it is better to tinker with indirectly measured parameters such as recharge?

But despite these reservations, properly compiled and rigidly constrained digital models are valuable tools with which to ask predictive 'what-if' questions. Given climate change, what if rainfall increases or decreases by this margin or that; what if the rainfall events become more intense and concentrated; what if that wellfield or this doubles its production; what if this river is diverted away from the aquifer, and so on.

The conceptual model

As we saw in Chapter 2, the first stage is to develop a **conceptual model** complete with estimates of throughput volumes and overall likely groundwater flow paths. This can be done from geological maps and other two-dimensional data, but nowadays can also be constructed in three dimensions using 3D visualization software. The latter is more expensive but will provide a more accurate basis for numerical modelling, and data can be transferred directly from the visualization software to the modelling software.

Groundwater investigation depends on the process of developing a conceptual flow model as a precursor to a mathematical model, which in complex aquifers may lead in turn to the development of a numerical approximation model. The assumptions made in the conceptual model depend heavily on the geological framework defining the aquifer, and if the conceptual model is inappropriate, then subsequent modelling

will also be incorrect. Paradoxically, the development of a robust conceptual model remains difficult, not least because this 3D paradigm is usually reduced to 2D plans and sections. 3D visualization software makes the model more robust and defensible and assists in demonstrating the hydraulics of the aquifer system.

There are two main types of modelling approach: digital analysis based on the groundwater flow equations, and process modelling. Process modelling, usually a spreadsheet model, allows input of variables according to the known processes that impact an aquifer and is especially useful when dealing with small-scale aquifers such as coastal dune systems. Some digital models also incorporate process features. Processes may include unsaturated vertical flow in the unsaturated zone, vegetation type, root depths and transpiration, field drainage, and so on.

Digital modelling

The basic inputs for the digital model are the hydrological variables, operational variables and the various boundary conditions, be they fixed head, no-flow, variable head, etc. The groundwater flow equation is applied to solve the flow analysis, commonly with the Laplace Transformation used to reduce the dimensions of the partial differential equation and simplify the computation of results.

Both initial conditions (heads at time (t) = 0) and boundary conditions (representing either the physical boundaries of the domain, or an approximation of the domain beyond that point) are needed. Often the initial conditions are supplied by a transient simulation, a corresponding steady-state simulation where the time derivative in the groundwater flow equation is set to 0.

Analytical methods of computing typically use the structure of mathematics to arrive at a simple, elegant solution. However, the required derivation for all but the simplest domain geometries can be quite complex (involving non-standard coordinates, **conformal mapping**, etc.). Analytical solutions also typically take the form of a simple equation that can give a quick answer based on a few parameters.

Numerical methods

There are two broad categories of numerical methods:
- gridded or discretized methods; and
- non-gridded or mesh-free methods.

In the common **finite difference method** and **finite element method** the domain is completely gridded into small elements. The analytical element method and the boundary integral equation method, also called the boundary element method, are only discretized at boundaries or along flow elements (line sinks, area sources, etc.); the majority of the domain is mesh-free.

Gridded methods

Gridded methods, like finite difference and finite element methods, solve the groundwater flow equation by breaking the area or domain into many small elements of squares, rectangles, triangles, blocks, or tetrahedra, and solving the flow equation for each element (all material properties are assumed constant or possibly linearly variable within an element), then linking together all the elements using conservation of mass across the boundaries between the elements. This results in a system which overall approximates the groundwater flow equation, but exactly matches the boundary conditions (the head or the **flux** is specified in the elements that intersect the boundaries). Finite differences are a way of representing continuous differential operators using discrete intervals (x and t). The finite difference methods are based on these. The model output is validated against observed data such as water level time series data.

MODFLOW is a well-known finite difference groundwater flow model that was developed by the United States Geological Survey as a modular simulation tool for modelling groundwater flow (Fig. 4.5). It is available free of charge, and is documented and distributed by the USGS. Many commercial products have grown up around it, providing graphical user interfaces to its input file-based interface, and typically incorporating pre- and post-processing of the user data. Many other models have been developed to work with MODFLOW input and output, making linked models that simulate several possible hydrological processes, e.g. flow and transport models, surface water and groundwater models and chemical reaction models.

Finite element programs are more flexible in design than most finite difference models, and use triangular elements as against block elements. There are several available programs: for example, SUTRA, a 2D or 3D density-dependent flow model by the USGS; Hydrus, a commercial unsaturated flow model; and FEFLOW, a commercial modelling environment for subsurface flow, solute and heat transport processes.

The **finite volume method** is a method for representing and evaluating partial differential equations as algebraic equations. Similar to the finite difference method, values are calculated at discrete places on a meshed geometry. Finite volume refers to the small volume surrounding each node point

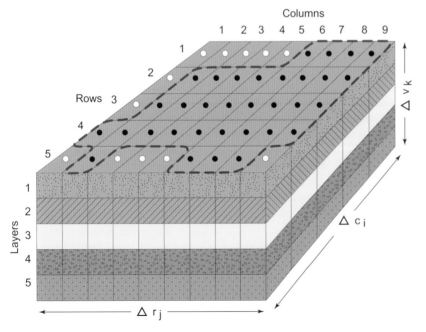

Figure 4.5 The 3D grid used in MODFLOW – adapted from MODFLOW Manual cover.

on a mesh. In the finite volume method, volume integrals in a partial differential equation containing a divergence term are converted to surface integrals. These terms are then evaluated as fluxes at the surfaces of each finite volume. These methods are conservative because the fluid entering a given volume is identical to that leaving the adjacent volume. Another advantage of the finite volume method is that it is easily formulated to allow for unstructured meshes. The method is used in many computational fluid dynamics packages.

Groundwater models can be one-dimensional or two-dimensional, but only two- and three-dimensional models can take into account the anisotropy of the aquifer with respect to the hydraulic conductivity, if this property varies in different directions. One-dimensional models can be used, for example, to evaluate vertical flow in a system of parallel horizontal layers.

Contrasting applications of groundwater modelling
The Palestine/West Bank conundrum – a politically sensitive model

Competition for water resources between Palestine and Israel is an ongoing cause of tension. The Western Aquifer Basin forms a major part of the complex, which is largely a karst, limestone system known as the West Bank Mountain Aquifer. The aquifer crops out and is recharged solely in the semi-arid uplands of the West Bank and groundwater flows west beneath Israel to discharge at the Yarqon and Nahal Taninim springs near the Mediterranean coast (Fig. 4.6). Annual recharge to the aquifer is not easy to quantify, but lies within the range 270×10^6 to $455 \times 10^6\,m^3$, but currently there is not sufficient reliable data to support a specific value of long-term average recharge. The resource is heavily exploited and abstraction is directly controlled and apportioned by Israel, between Israel and the West Bank.

The key to equitable apportionment is determination of the long-term average recharge to the groundwater system, but this also requires definition of the eastern boundary of the basin to confirm the recharge area. Calculations include empirical formulae and process-based models that may constrain the best estimates, provided that there is appropriate, ongoing monitoring. Improved understanding can then be fed back into the model. All approaches suffer from uncertainties, which hinder communication with politicians and managers, who tend to interpret the

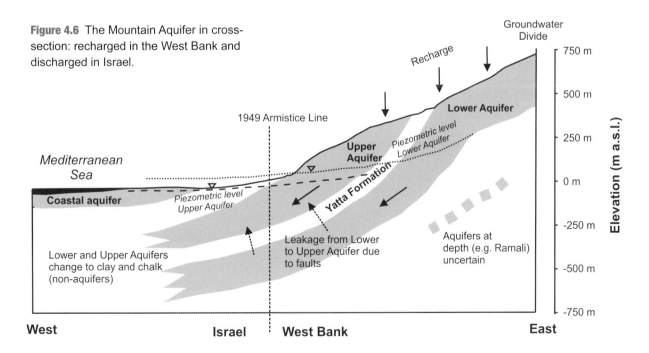

Figure 4.6 The Mountain Aquifer in cross-section: recharged in the West Bank and discharged in Israel.

range of values available as a lack of knowledge regarding the working of the groundwater system.

Dumfries basin, southwest Scotland – a finite element model in support of groundwater management

The Dumfries Basin aquifer, some 10 by 10 km in extent, supports groundwater abstraction for public supply, agriculture and industry. Abstraction is concentrated in the western part of the basin, where falling groundwater levels and deteriorating water quality both reflect the effects of intense pumping. There are two bedrock units: a predominantly breccia/coarse sandstone sequence in the west, interfingering with a predominantly sandstone sequence in the northeast and east. The basin is bounded by weakly permeable Lower Palaeozoic rocks, and is largely concealed by a variety of superficial deposits. Surface water flows onto the basin from the surrounding catchments via the River Nith and the Lochar

Water and their respective tributaries. Direct rainfall recharge occurs via superficial sands and gravels, especially in the north, and discharge is predominantly to the rivers in the central area of the basin rather than to the sea, which lies to the south. There are two main aquifer types within the basin: the high-transmissivity western sector underlain by a fracture-flow system with younger water, active recharge and a high nitrate content, compared to the east, where groundwater residence times are longer and the storage capacity is higher.

Numerical simulation was used to establish the water balance for the basin to support development of new abstraction boreholes for public supply (Fig. 4.7). The model was also applied to support the conceptual flow model of the basin and confirm the no-flow barriers identified in that model. The relationship between the River Nith and the Permian sandstone aquifers was also quantified in the simulation model (Fig. 4.8).

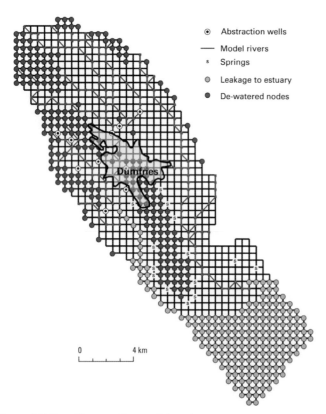

Figure 4.7 Dumfries aquifer model nodes.

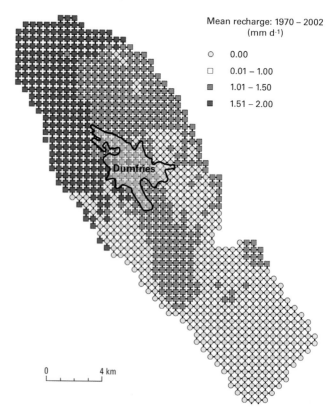

Figure 4.8 Simulated recharge estimates in output from the Dumfries model.

The key model parameters are:

- Effective rainfall – 436 mm per year;
- recharge is constrained by diversion to field drains and runoff induced over clayey areas, and the total average annual rainfall recharge volume is 25 Mm³ year, equivalent to 68 000 m³ per day;
- Surface water ingress to the aquifer is likely to be small;
- Loss from sewers and water mains, irrigation returns and other discharges to the ground is small;
- Groundwater discharge to the River Nith – potential for 5000 m³ per day per km of river.
- There is also some groundwater discharge to a small lake;
- Groundwater discharge directly to the sea is small;
- Groundwater abstraction is between 15 000 and 30 000 m³ per day.

A key outcome of the model was identification of areas of greatest uncertainty. This targeted field investigations that were then designed to look further at the relationship between groundwater in the sandstone aquifer and water in the River Nith, and at discharge into the tidal estuary of the Nith at the southern end of the aquifer.

Vowchurch model of shallow valley alluvial fill – an unnecessary model

Analysis of a granular valley-fill alluvial aquifer, partly overlain by head deposits and coarse till material at Vowchurch in Herefordshire, UK, showed that modelling is not always essential to solving problems. The shallow aquifer system occupies a 200 to 300 m width of the River Dore valley and is generally no thicker than 5 m. The main sand and gravel aquifer has a permeability of between 50 and 90 m day⁻¹ determined from testing at an abstraction borehole. The borehole is 300 m downstream from a proposed crossing of the valley by a 1.5 m diameter gas pipeline. The questions were asked: what will be the impact of trench dewatering on the pumping station during dewatering and pipe emplacement, and what will be the long-term damming effect of the pipe itself on groundwater flow down the length of the valley?

Simple Darcian analysis of the likely depletion at the pumping station by the proposed dewatering works indicated that water levels at the pumping station were likely to fall by about 0.5 m. The greatest impact would be on reduced base-flow into the River Dore down gradient from the pumping station. Likewise, the reduction in transmissivity caused by removing a small segment of the saturated aquifer where the pipeline would lie showed that there was likely to be a minor damming effect of only a few centimetres up gradient of the pipeline. A comprehensive programme of radial flow analysis of the original pumping test data, distributed recharge calculations and groundwater flow modelling served to confirm these calculations. The additional cost of this work was some £50,000; the benefit was a substantial groundwater flow model, but based on the same scanty database as the simple Darcian calculations, with basically the same output. In this case the Darcian estimate was enough to help the pipeline engineers and other stakeholders, and modelling was unwarranted.

5   Boreholes and test pumping

Boreholes, wells and galleries

For many centuries groundwater has been accessed by digging wells, drilling boreholes and the use of water pumps (Fig. 5.1). A Hy-Mac, or similar excavator, can attain 5 m depth in soft material; a modern air percussion drilling machine can drill to 120 m depth in one shift. The days of digging a 120 m deep well in hard volcanic rock by hand, as once they did in the Middle East, are now thankfully over.

Shallow, unconsolidated drift deposits can be dug to beneath the water table, and a shaft of preformed concrete rings let in before backfilling the **annulus** with coarse granular material. Stones can be placed in the joints to part the rings and allow water to enter the shaft. Coarser backfill should be placed adjacent to these openings, and cobbles can be laid on the bottom of the shaft to prevent upwelling of fine material.

Figure 5.1 Borehole headworks and pump house, Matui Village, Dodoma District, Tanzania. Note the derrick used for lifting the pump for maintenance. Image credit: Jeff Davie.

Groundwater inflow from the base of the shaft or well of diameter D m during pumping at q m³ d⁻¹ (but excluding any inflow from the open ring joints) allows a simple estimate of hydraulic conductivity of the aquifer to be made from the relationship:

$$q = 2.75\,DkH$$

where k is hydraulic conductivity (m day⁻¹) and H is the drawdown (m).

Harder rock and deeper penetration require the use of drilled boreholes. The completed diameter of a borehole should be adequate to house an electric **submersible turbine pump** of suitable capacity yet leave sufficient annulus around the pump bowl not to starve water flowing past it towards the pump intake. The borehole may be lined with mild steel, or more commonly, various types of plastic. Stainless steel **well screen** or slotted pipe are placed at intervals to allow groundwater to enter from the more productive horizons. These zones are usually identified with the aid of **geophysical logs** run within the drilled hole prior to running-in the borehole casing (Fig. 5.2).

Types of well screen

The most common slot width for plastic screen is 2 mm, although numerous other sizes are available. Slotted screen is also available as textile-wrapped screen, allowing a finer-grained formation to be screened. Careful slot width selection is required for stainless steel wire-wrapped screens, which have the advantage of offering a greater open area to the aquifer (greater hydraulic efficiency) and enable even finer-grained unconsolidated zones to be exploited. Sieve analysis of unconsolidated sand samples prescribes the slot width, which is normally selected so that 50% to 60% of the sample can pass through the screen. Development of the screened sections is undertaken by washing and surging the borehole, sometimes using a vigorous air lift pump, sometimes by jetting water into the screen, and sometimes by swabbing

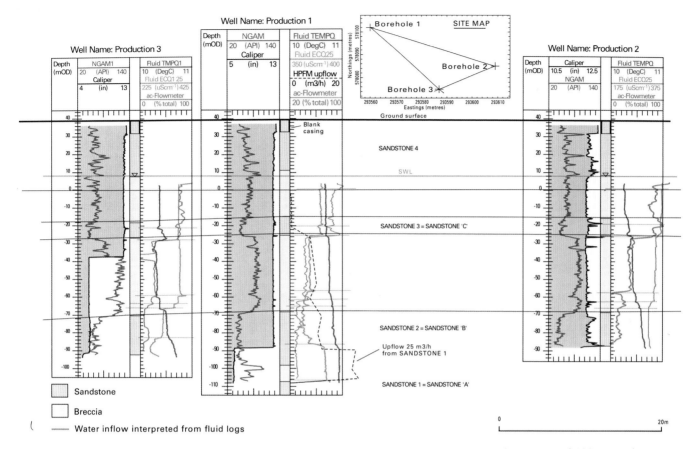

Figure 5.2 Geophysical logs run in three boreholes in Permian sandstone and breccia, Dumfriesshire: NGAM, natural gamma; Caliper, borehole diameter; TMPQ, water temperature; ECQ, fluid specific electrical conductance; flowmeter, natural flow in borehole.

the screen sections. These actions allow fines to pass through the screen into the borehole to be removed by bailer or pump. This produces a graded filter pack leading from the borehole screen into the aquifer.

A graded filter pack can be employed by placing a rounded and well-sorted filter medium between the screen and formation. The practicalities of emplacement require a minimum 50 mm annulus. The pack material should be of a grain size a little larger than that of the formation. **Gravel packs** are particularly useful in fine-grained unconsolidated sands. The median grain size of the pack should be such that it will retain about 60% of the formation grains, and the screen slot size should be such that it will retain about 90% of the pack. A

formation stabilizer, however, is a medium- to coarse-grained rounded gravel placed behind the screen and designed to hold the formation from sloughing against the borehole casing. Its grain size is not critical. It is used in formations such as consolidated but friable sandstones and other strata that would not stand indefinitely in an open hole. Open-hole completion may be suitable for some hard limestones and unweathered volcanic rocks.

Good practice

It is good practice to have a sump of plain casing, capped at the bottom, and placed below the lowest screen section. This allows debris that enters the borehole to accumulate clear of the screen section. The pump should ideally be placed within a section of plain casing, but this is not always possible. The optimum pump setting depends on the likely **drawdown**. A cut-out electrode is useful above the pump intake to protect the pump from drawing air. Other pump types are available, but the submersible turbine pump is the most commonly used pump for most applications.

An infiltration gallery, somewhat akin to a sub-horizontal well, is a useful means of abstracting from a shallow unconsolidated aquifer. A trench should be dug to a level just deeper than the water table, and a slotted pipe let into it and covered in coarse granular material. The trench is then backfilled to ground level. The pipe is set at a slight angle and drains down into a sump at the bottom of a shaft from which groundwater can be pumped. The purpose of the gallery is to distribute the abstraction of water along its length and so reduce the overall drawdown during abstraction.

An observation well or piezometer is a narrow diameter borehole screened below the water table in which water level or head can be measured.

It is normal practice to place a sanitary seal around the upper section of casing to protect the annulus from surface contamination (Fig. 5.3). Ordinary cement is an effective seal, tamped down to about 1 or 2 metres depth to rest on top of the formation stabilizer. A concrete plinth at surface is a neat way of finishing off, and the hole should be padlocked to prevent unwanted access.

Where do I drill for water?

First of all, we need to understand the geology of the aquifer. Only then will we know enough about the hydrogeology to identify optimum locations for drilling groundwater abstraction points.

When faced with a groundwater development project in an aquifer, large or small, the first thing to do is to gather all the pertinent data that may be available. A geological map and cross-sections can be used to create a three-dimensional image of the aquifer shown draped beneath the topography. Borehole data may be available; in the UK this is available free of charge from the BGS website. Reports and even published papers may provide further insight into the current understanding of the working of the aquifer. The next task is to develop a conceptual flow model and try to estimate the components of the water budget. This allows us to identify

Figure 5.3 Malawi: a sophisticated, but damaged, sanitary seal allowing runoff from the well head to water a garden; note the table for washing clothes beyond the borehole pump. Image credit: Jeff Davies.

recharge and discharge zones, as well as giving us some idea of the pumping rates that the aquifer can sustain. If possible, develop a simple flow-net in order to identify those areas of likely higher permeability as well as directions of groundwater flow.

So where is the best place for my borehole? Groundwater is generally deeper beneath more elevated ground and interfluves than it is beneath the valleys. A shallow borehole in a valley bottom is cheaper to drill and maintain than a deeper one on the interfluve. Besides, fracturing and weathering tends to be greater in valleys than on hilltops, and this promotes higher storage capacity and greater permeability beneath the lower-lying land.

There are a number of local features too, and these are especially useful where no maps exist and only aerial photos are available to guide the prospector. Termite hills generally indicate moisture, and some trees, for example eucalyptus and willow, almost always have a root system that passes down to the **capillary fringe** of the water table. There may be springs and seepages to guide borehole siting. There may also be obvious pollution sources in the area that need to be distanced from any new groundwater source.

If still uncertain where best to site the borehole, surface geophysics can be helpful. A common technique is electrical resistivity surveying, a rapid survey technique that often highlights the depth to the water table and the overall salinity of the groundwater. However, as with all geophysics, it is essential to have a known data element, and it is useful to include a borehole or spring source within the survey to provide fixed **data truthing** for depth to the water table and for water salinity and its electrical conductance.

Borehole drilling, contracts and pitfalls

There are two basic types of drilling techniques: percussion and rotary.

Percussion drilling

Traditional percussion or cable tool drilling is the repeated lifting and dropping of a heavy set of drilling tools, essentially a wedge-shaped steel bit, onto the bottom of the hole. The bit breaks and crushes the rock into fragments in consolidated ground and loosens the formation in unconsolidated material. Mixed with water, which may be added in a dry hole above the water table, the resultant slurry can be lifted to the surface with a bailer. In very soft ground a sand bailer or shell is used to drill. Particles from the slurry are used to describe the lithology of the strata penetrated to form a geological log of the borehole. The up and down movement of the bit is imparted by the cable, either by the driller's use of a clutch and brake or via a spudding beam that imparts the up and down motion and is driven by a crank connected to a motor.

The efficiency of percussion drilling in consolidated formations depends on the resistance of the rock, the dip of the rock structure, the weight of the drill tools, the length and rate of the stroke, and the diameter, sharpness and shape of the bit. Clearance between the tool string and the borehole wall, as well as the density and depth of accumulated slurry at the bottom of the hole, are also important. Efficiency in unconsolidated sediments is more dependent on the presence of cobbles and boulders, which impede the process, than on the mode of drilling. It is usually necessary to drive casing behind the bit in order to keep the hole open.

Rotary drilling

Rotary drilling allows faster penetration rates in most formations (Figs 5.4 and 5.5). The drilling bit is attached to the end of a drill pipe, which is rotated so that a toothed set of wheels at the lower end of the pipe can grind the bottom of the hole by a process of cutting and chipping to allow penetration of hard materials. Different drill bits are used depending on the nature of the formation to be drilled. Drilling fluid is circulated down the drill pipe and back through the annulus between drill pipe and borehole wall. This serves to cool and lubricate the drill bit, to remove debris from the hole and support the walls of the shaft. The rotation of the bit is generally imparted directly from the drilling rig to the drill pipe. However, a technique called rotary percussion uses compressed air in the drill pipe to drive the bit in a percussive rotation and is particularly useful in exceptionally hard formations (Fig. 5.6).

Figure 5.4 Air percussion rotary rig working in Ghana. Image credit: Jeff Davies.

44

Figure 5.5 A Gemco mud flush rotary drilling rig in Fiji – note the drilling bit, left foreground. Image credit: Jeff Davies.

Drilling fluids

Drilling fluid may be water, compressed air, bentonite mud or a viscous organic fluid designed to reduce to the viscosity of water when chemically treated. The viscosity and density are critical for the removal of the drill chippings. The drill chippings should be collected at intervals and recorded in a log of the formation material penetrated by the borehole. It is also useful to record penetration rate, as this will change

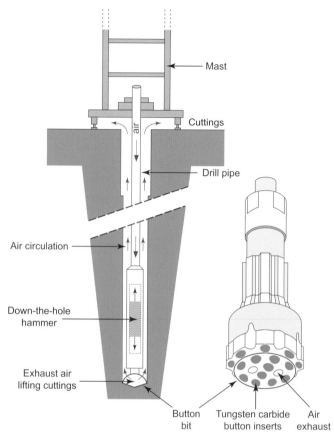

Figure 5.6 The air-percussion drilling process. Image credit: Jeff Davies.

with different formations and provide the precise depth to formation changes. Depth to the bit is measured from the **rotary table** as the number of drill pipes (usually 3 m long) that are in use, plus the drill stem, minus the height to the rotary table above ground.

The selection of an appropriate drilling fluid depends on the nature of the formation to be drilled. Bentonite-based muds cannot readily be removed from fine-grained material adjacent to the borehole, even though a filter cake will form on the borehole wall to inhibit mud ingress. The mud filter cake should be thinned after completion of drilling by circulating water for several hours before inserting the casing and gravel pack. If this is not done it is generally very difficult to remove residual bentonite from around the borehole during

Figure 5.7 Air lift surging in a newly completed borehole in the Tihama, Yemen.

subsequent washing and well development. Water and air are by far the best fluids for circulation, but it may be necessary to use more viscous and dense fluids in difficult formations. A thorough programme of borehole cleaning is always worthwhile and should include airlift surging, scouring, jetting, the use of softeners such as Calgon, and even acid in limestone formations (Fig. 5.7).

Coring

It is sometimes necessary to retrieve solid core samples from the formation for laboratory testing and lithological description. This can be done with a conventional rotary core barrel, although care is needed to select equipment that is suitable both for the drilling rig and the ground to be cored. Percussion drilling allows samples to be taken using a U100 percussion barrel, but only in unconsolidated material.

The drilling contract

In the UK, the tender document for a drilling contract should follow the procedure set out by the Institution of Civil Engineers Standard Contract Form. The significant parts of the contract are the Specification for the work to be done, i.e. of the boreholes, their dimensions and their means of completion, and the Bill of Quantities. It is good practice to allow contractors as much room as possible in order to give them the choice of drilling technique, a range of sizes, tolerances and materials. Prognosis of the probable ground conditions must be made carefully and avoid being too specific if there is any doubt about the geology. Indeed, the contractor would have grounds for a claim if he tooled up the rig for Carboniferous Limestone and found he had to drill London Clay!

Unlike other civil engineering contracts, the drilling contract should be flexible because the complete job cannot be surveyed before commencement, and quantities cannot be estimated precisely. A good contract document and good supervision of the contractor should allow some element of give and take, and contractual difficulties must never be allowed to sour relationships on site. Should contractual difficulties arise, these should be dealt with in the office and not by driller and hydrogeologist on site.

Test pumping a borehole

The practicalities of carrying out a **test pumping** exercise require a little planning. When a new borehole is drilled, it will be surged and over-pumped to get some idea of the optimum rate of pumping – that is, the maximum pumping rate that will not expose the pump over the duration of the test. Once the pump has been installed it is customary to test it for half an hour to check the likely maximum yield possible for the test. The pumping phase of the test may be three days pumping, seven days or longer, and the recovery monitoring phase following shut down of the pump, when the water level rises again, will be of equal duration. The sustainability of supply can only be realized once the source has been commissioned; however, the longer the duration of test pumping, the more likely that the outcome yield will be sustainable.

Figure 5.8 Bangladesh: test pumping a deep borehole in Bangladesh, discharging into a lined channel to prevent water recycling back to the aquifer. Image credit: Jeff Davies.

It is important to install a discharge main to remove the water from the site so that it cannot recirculate back into the aquifer if it is unconfined (Fig. 5.8). The main needs to include either a flow volume metre, a manometer or a 'v' notch weir so that the amount and rate of water discharge can be monitored. A well head sample tap is useful if samples are needed for chemical analyses. A submersible turbine pump is the most flexible with which to adjust flow rates using a surface valve on the discharge line. A three-phase electricity supply is needed for the pump.

Transducers should be set in the pumping borehole and all the nearby observation boreholes to record water-level decline – the drawdown of the water level over time. The water level at start and finish of each transducer needs to be measured with a dipper. The time series need to be coordinated so that each record notes exactly when the pump starts and stops. On completion of the pumping phase the transducers are left in the boreholes to record the recovery of the water levels. These are important data as they are not affected by variations in flow rate caused, for example, by changes in the efficiency of the power supply generator with air temperature night and day.

Radial flow to a well and test pumping techniques

Test pumping enables the formation constants, permeability and storativity and also **well efficiency**, to be determined. The aquifer is stimulated by pumping at a constant rate approaching the maximum capacity of the well, and observing the response or drawdown in the pumped borehole and in nearby observation boreholes or **piezometers** (Fig. 5.9). Well efficiency is determined by running a stepped test at incremental discharge rates with each of four or five steps pumped for two hours and the drawdown recorded in the pumped well against time. The incremental drawdown will increase with each step, with increasing rates of drawdown at the higher discharge rates reflecting reduced efficiency caused by turbulence at the well periphery as groundwater enters the borehole column (Fig. 5.10).

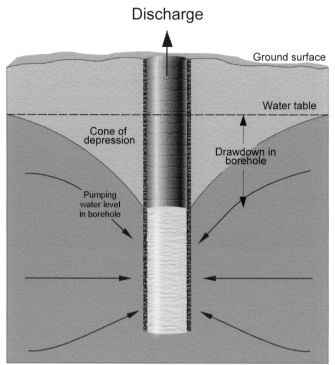

Discharge

Ground surface

Water table

Cone of
depression

Drawdown in
borehole

Pumping
water level
in borehole

Figure 5.9 The effect of a pumping borehole on the water table in an unconfined aquifer. Image credit: UK Groundwater Forum.

Constant rate test-pumping data are interpreted by using an analytical model of aquifer flow to match the observed data, and by applying best fit values of permeability and storativity. In more complex cases, a numerical model may be used to analyse the results of a test.

An appropriate model or solution to the groundwater flow equation must be chosen to fit the observed time series data. There are many different choices of models, depending on what factors are deemed important, including:

- leaky **aquitards**;
- unconfined flow (delayed yield);
- partial penetration of the pumping and monitoring wells;
- finite borehole radius;
- dual porosity (typically in fractured rock);
- anisotropic aquifers;
- heterogeneous aquifers;
- finite aquifers (the effects of physical boundaries are seen in the test); and
- combinations of the above situations.

Nearly all aquifer test solution methods are based on the **Theis equation**. This is given below as just one example of the formulae in common usage during hydrogeological investigations. More detailed explanations of the mathematics of groundwater flow are available in textbooks (see recommended further reading on page 108).

47

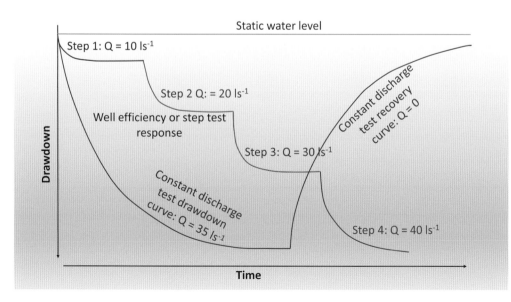

Static water level

Step 1: Q = 10 ls⁻¹

Step 2 Q: = 20 ls⁻¹

Well efficiency or step test
response

Constant discharge
test recovery
curve: Q = 0

Step 3: Q = 30 ls⁻¹

Constant discharge
test drawdown
curve: Q = 35 ls⁻¹

Step 4: Q = 40 ls⁻¹

Drawdown

Time

Figure 5.10 Idealized step test discharge borehole efficiency test and constant yield test drawdown responses with indicative yields (Q).

The Theis equation is:

$$s = \frac{Q}{4\pi T} W(u)$$

$$u = \frac{r^2 S}{4Tt}$$

Where s is the drawdown, u is a dimensionless time parameter, Q is the discharge (pumping) rate, T and S are the transmissivity (permeability times aquifer thickness) and storativity of the aquifer around the well, r is the distance from the pumping well to the point where the drawdown was observed, t is the time since pumping began, and $W(u)$ is the well function, which is an exponential integral.

The Theis solution

The Theis solution is built upon simplifying assumptions. Other methods relax at least one of the assumptions on which the Theis solution is built, and can achieve a more flexible result. It is not uncommon to apply several different methods to analyse one set of test data and compare the results.

The Theis solution is based on the following assumptions:

- The flow in the aquifer is adequately described by Darcy's Law;
- homogeneous, isotropic, confined aquifer;
- well or borehole is fully penetrating (open to the entire thickness (b) of aquifer);
- the well has zero radius (it is approximated as a vertical line) so that no water can be stored in the well;
- the well has a constant pumping rate Q;
- the head loss over the well screen is negligible;
- aquifer is infinite in radial extent;
- horizontal (not sloping), flat, impermeable (non-leaky) top and bottom boundaries of aquifer;
- groundwater flow is horizontal;
- no other wells or long-term changes in regional water levels (all changes in potentiometric surface are the result of the pumping well alone).

Even though it is rare for all these assumptions to be met, the solution is still useful.

Theis, or any other analytical equation, finds the average T and S values near a pumping borehole from drawdown data collected during the aquifer test or test pumping exercise. A simple form of inverse modelling is applied: the drawdown is measured in the well, distance to observation point, elapsed time since pumping started or stopped, and discharge rate are observed, and values of transmissivity and storage that may reproduce the measured data are put into the equation until a best fit is achieved. It is not necessary to understand the mathematics behind the analyses, but it is necessary to be able to identify which formulae are appropriate, and under which conditions. Always check that the derived values of permeability and storativity fit the range of values likely for the aquifer material. If the values do not fit, then the analysis method is probably inappropriate or there has been some error in gathering or handling of data.

Some useful concepts and some concerns

Before looking at any borehole performance data in an aquifer or a whole aquifer system, work out whether they penetrate unconfined or confined aquifers, according to the geological setting. This is important because the data will be quite different for the two types of aquifer, and the performance of the aquifers will be quite different as well. Next, it is useful to allocate a predicted permeability and storativity value to the aquifer, based on typical values of aquifers elsewhere that share similar lithology and morphology. Armed with these estimates, guess what sort of yields are likely from the boreholes. Now look at the data available from the boreholes. If this is not of the same order of magnitude as your own estimates, treat the borehole data with great caution and plan to gather your own field and borehole test data in due course. So often in the past people have adopted data without checking its validity, loaded it into a time-consuming and expensive digital model, and produced nonsense, albeit convincing-looking nonsense!

Comparisons of the performance of one borehole with another in the same or similar aquifers can be made with the specific capacity value. **Specific capacity** is simply the discharge rate of a borehole divided by the drawdown at an assumed steady state; that is, the drawdown remains reasonably constant while pumping continues, commonly at an elapsed time of 24 hours since the pump was switched on. In low-yielding aquifers such as the crystalline basement aquifer of Subsaharan Africa, a shorter elapsed time may be used such as three hours or eight hours. However, it is important that the specific capacity is quoted with the pumping time, e.g. 2.0 l/s/m for eight hours' pumping. The higher the value of specific capacity, the more productive the borehole will be.

Values can be plotted on a map and even contoured to identify the better-yielding parts of an aquifer.

In some areas the only quantitative data that may be available is borehole specific capacity. Do look at the original data to check if the values really are steady state values that can be compared one to another.

Many borehole records will provide a maximum pumping value or 'safe' long-term yield. Again, look at the data and check where this value comes from. More often than not it was the maximum yield attainable with the equipment available during the original tests on the borehole. It may have nothing to do with safe yield.

Records often report values for transmissivity rather than permeability. However, transmissivity is not a formation constant, being a product of effective aquifer thickness and permeability (k). In shallow unconfined aquifers the drawdown caused by pumping reduces the thickness of saturated aquifer around the well so that the transmissivity changes from k(thickness d_1) at the time pumping started to $k(d_t$ – drawdown) as drawdown increases. In very thin aquifers the difference may account for a 50% decrease in transmissivity. The other problem with transmissivity is that it varies with the geometry of the aquifer; it is lowest in shallow areas and highest in thicker areas.

Finally, many textbooks explain that almost all groundwater flow analysis requires very specific conditions to be met. These include aquifers of infinite extent, totally isotropic and homogeneous, and a variety of other constraints that will never be met in the real natural world. For the most part the errors caused by not meeting these conditions have only a minor influence on the results of the analysis and only need to be met as best they can.

6   Groundwater management

Overview

The conventional vision for groundwater management, based broadly on welfare economics and integrated water resources consumption, has four basic components:

- Water should be treated as both an economic and a social resource and it should be priced accordingly;
- That 'rights' should be specified and enforced according to agreed social, economic and environmental priorities;
- That the concept 'the polluter pays' is laudable;
- Surface and groundwater should be considered as an integrated and holistic resource.

These four parts of the management vision readily translate into an overall management prescription:

- Public ownership of groundwater needs to be legislated, with state or government granting rights of use;
- An allocation/licensing system needs to be developed, based on sustainable yield assessments, with priority given to domestic use;
- Groundwater pricing needs to be introduced, based on the principle that water is an economic as well as a social good. Users in poor communities need to pay at least part of the economic cost of groundwater including its marginal costs of supply;
- A groundwater pollution protection strategy needs to be developed and implemented;
- The environmental services provided by groundwater, including baseflow to rivers and wetlands, needs to be recognized and protected within an integrated planning framework.

These ideals, however, are hard to achieve in many countries, particularly by the poorer nations. Lack of data and thus lack of understanding of the groundwater resource potential is a key constraint. A particular problem in Africa and Asia is the difficulty of coordinating and controlling the large numbers of groundwater users in rural farming communities. Institutional capacity may also be a major constraint, particularly with regard to customary usage of groundwater prior to enactment of new legislation.

Concept of 'safe yield'

In most water-scarce areas, burgeoning population and socio-economic development are causing depletion of groundwater resources, seawater intrusion, and reduction in baseflow to rivers. In order to satisfy the environmental and hydrological requirements, optimization approaches can be applied to determine groundwater use strategies. This is essential in order to manage and use a resource efficiently. Controlling withdrawals from aquifers without exceeding the 'safe yields' is an emerging water supply issue.

As groundwater is considered to be a renewable natural resource, only a certain quantity of water may be safely withdrawn from a basin or aquifer. The maximum quantity of groundwater that can be abstracted from boreholes and wells while still maintaining an unimpaired supply depends upon the **safe yield**.

The safe yield of a groundwater basin, or of just one aquifer unit, is the amount of groundwater that can be abstracted, usually annually, without producing an 'undesired effect'. Any abstraction in excess of the safe yield is **overdraft**. This is a seemingly simple concept – only so much water enters the basin, so no more than that amount can be removed by abstraction. But a number of issues need to be considered in addition, not least the economics and viability of pumping, the quality of the water, the management strategy for the aquifer, and any existing **water rights**. And just what is meant by 'undesired effects'?

Avoiding 'undesired effects'

This will mean different things to different people and to different stakeholders. Primarily, it means safeguarding the aquatic system that has developed at the surface, draining from a discharge zone of an aquifer. It is essential that sufficient water is allowed to discharge to maintain the aquatic system below the groundwater discharge zone, and also that enough water is discharged to satisfy riparian rights that may exist for abstraction of the surface water flow. Equally, the equilibrium of the system should not be damaged in such a way that deeper, more mineralized or brackish water is drawn up to the pumps, or that pumping water levels decline to increase costs of pumping.

Changes that can occur within a basin include those resulting from climate change, changes to water rights and management, increasing depth of pumping and associated cost implications, as well as deterioration in the quality of the water abstracted. Returns to an unconfined aquifer from, for example, leaking water mains and sewers, and spray irrigation, can be deducted from total abstraction. All of these factors need to be considered from time to time to adjust the safe yield value. Increasing depth of pumping and deteriorating water quality both imply some form of overdraft. But changes might also occur in neighbouring basins or aquifers, and these too might impact the safe yield of the basin.

Calculating safe yield

Various formulae have been published that try to formalize the calculation of safe yield. None are robust and none are universal. It is preferable to provide sufficient understanding and data for the development of a digital model of the groundwater system that will run 'what-if' scenarios to check the safe yield value. The actual value may vary from perhaps only 30% of groundwater throughflow in a large aquifer system to 90% in, say, the Chalk Brighton Block of southern England, which is bordered on two sides by salt-water tidal rivers, on the third by the sea and on the fourth by the scarp slope of the Chalk outcrop; i.e. in this case there is very little discharge to land on which aquatic systems or riparian users are dependent.

There are a number of subtle effects that influence safe yield. For example, pumping from an abstraction borehole near a surface water course may intercept groundwater that would otherwise flow into a river as baseflow. This reduces the supply of water that keeps the river flowing in the dry season. The effect of pumping a borehole is to lower the groundwater table surrounding it into a **cone of depression**. This cone changes the slope of the groundwater table locally, which can create a gradient from the nearby river toward the borehole, resulting in either:

- intercepting groundwater baseflow that would otherwise discharge into the river, resulting in reduced surface water flow; or

- inducing infiltration, thus causing surface water from the river or stream to flow from there through the ground and into the borehole, also resulting in reduced surface flow.

These effects are sometimes exacerbated by high demand for groundwater during the summer and early autumn. Seasonal pumping occurs exactly when the natural flow system is at its most critical stage – low seasonal flows coupled with the growing season.

Optimum aquifer usage

The goal for environmentally sound and sustainable development of water resources is to develop and manage them so that the resource base is maintained and enhanced over the long term. Groundwater development begins typically with a few pumping wells, and initially the groundwater management practice is geared to facilitate usage and development. As development progresses with more and more drilled wells scattered over the basin, issues such as **over-exploitation**, equitable sharing of water and degradation of water quality become apparent. The emphasis of groundwater management practice has then to be changed so that the available resource is utilized in an efficient, sustainable and equitable manner that will contribute to the economic and social well-being of the broader community.

Sustainable groundwater development depends on the understanding of processes in the aquifer system, quantitative and qualitative monitoring of the resource and its interaction with land and surface water environments. Key principles to consider when developing a sustainable groundwater resource are:

- long-term conservation of the resource;
- protection of groundwater quality from significant degradation; and
- consideration of resulting environmental impacts.

Aquifers that are exploited for public water supply, industrial use, power generation and agriculture need careful management to ensure that supplies are able to cope with future increases in demand and are resilient to a changing climate. Groundwater baseflow to rivers and streams and support to some wetlands still needs to be safeguarded; groundwater-dependent terrestrial ecosystems are, of course, important habitats. Groundwater is easily polluted and can be difficult and expensive to remediate, therefore it is essential to manage discharges or other activities that can cause pollution.

Water quality legislation

In the United Kingdom, the Water Act, 2003 was implemented not only to streamline abstraction licensing, but also to provide greater focus on water conservation and protection of the water environment. Some activities that were previously exempt from licensing, such as mine and quarry dewatering and trickle irrigation, now require a licence.

European legislation has also had a significant impact on the protection and management of groundwater, the most significant of which has been the Groundwater and Groundwater 'Daughter' directives, the Habitats Directive and the Water Framework Directive. Between 2001 and 2008 the Environment Agency of England and Wales also developed Catchment Abstraction Management Strategies (CAMS) to promote groundwater and surface water resource management. These consider recharge, the water requirements of the environment, including groundwater baseflow to support flow in rivers, and the amount of water licensed for abstraction. CAMS highlight areas of water availability for future licensing and areas that are already over-licensed. However, they only cover the main aquifers and smaller aquifers that have a large number of groundwater abstractions. Aquifer designations in England and Wales – either Principal or Secondary Aquifers – reflect the importance of the aquifer both as a resource and in its role in supporting surface water flow and wetland ecosystems. Principal aquifers consist of bedrock aquifers that can supply water on a strategic scale. Secondary aquifers include a wide range of rock types or drift deposits with an equally wide range of water permeability and storage potential.

To complement abstraction licensing and the permitting regimes that control activities that may pollute and **derogate** groundwater, the regulators in England and Wales have developed a number of assessment tools. These allow a risk-based approach to be adopted, focusing on aquifers that:

- are exploited for public water supply;
- have known problems with over-abstraction;
- have been identified at 'poor' status under the Water Framework Directive;
- require information to respond to future pressures such as climate change.

So-called River Basin Management Plans have also been developed to identify measures to achieve Water Framework Directive requirements for all water bodies.

Source Protection Zones are used to protect abstractions used for public water supply and other forms of distribution to the public, such as mineral and bottled water plants, breweries or commercial food and drink production sites (Fig. 6.1). These zones show the areas of groundwater within which there is particular sensitivity to pollution risks due to the proximity of a potable source. Whereas protection zones around smaller sources can be delineated with analytical or numerical modelling techniques, in areas of karstic, non-Darcian flow they are best defined by tracer testing and field examination.

Protection of the resource is provided by **Nitrate Vulnerable Zones**. These delineate areas of agricultural nitrate pollution to meet the requirement of the EU Nitrate Directive (91/676/

Figure 6.1 'Water collection area' notice in the Netherlands indicates a protection zone for a public supply groundwater abstraction source. Image credit: Shutterstock/ Marieke Kramer.

EEC). Where groundwater quality demonstrates increasing nitrate trends above the trigger of 11.3 mg-N l⁻¹, the **drinking water standard**, further investigation, for example of land management practices, is applied to refine the catchment areas. The zones are reviewed every four years and currently cover around 70% of the land area in England.

Conjunctive use and river augmentation

Conjunctive use is the combined abstraction of surface water resources and groundwater, in a unified way, in order to optimize the use of the resources and minimize adverse effects of using just surface water or just groundwater. This concept exploits the storage capacity of an aquifer and the ease of transport of water by a river. The aquifer is used to store surface water when there is an excess of it, for example, in the rainy season or in winter. The river is used to transport water from the aquifer to where it is needed when the river discharge is low, as often happens in summer. Conjunctive use can also reduce abstraction from rivers when the discharge is low by preferentially using groundwater.

Managed aquifer recharge

The storage of excess surface water underground in an aquifer is a type of conjunctive use called **managed aquifer recharge**. This makes the most of excess water by directing it into the ground where it can be stored for future use. Underground storage has many advantages over surface storage: no land is taken up by reservoirs, there is no evaporation loss, and capital costs are much lower. However, managed aquifer recharge is not a simple process, and complex issues need to be overcome. It cannot absorb large volumes of flood water in a short time. It involves transferring water from the surface to underground, either by dispersing it over the surface to increase infiltration, or through aquifer injection wells.

Surface dispersal involves diverting the water into an unlined canal or shallow lagoon over permeable sediments or rock so that the water can percolate downwards into the aquifer. It works best in areas with highly permeable soils and unconfined aquifers, and where land is inexpensive.

Aquifer storage and recovery schemes use the same borehole to inject and recover water. While most systems are designed to store water during the wet season and recover it during the following dry season, some are established for water banking, where recovery may not take place for several years.

Artificial recharge and recovery take place in the Lee

Valley, north of London. Here water is often in short supply in summer, and groundwater from the underlying Chalk and Basal Sands aquifer cannot cope with seasonal demand. In the 1970s an artificial recharge scheme using injection wells was started into this aquifer. The water used for recharge is from the rivers Thames and Lee at times of excess flow in winter. It is treated to drinking water standards before it is injected into the aquifer (when spare treatment capacity is available) so that there is no danger of polluting the existing good-quality groundwater (Fig. 6.2 – summer and winter). The scheme recharges the aquifer over an area of 50 km² and provides

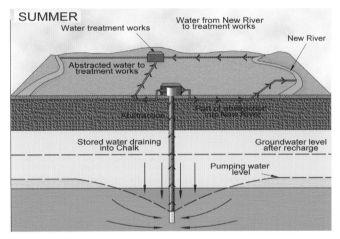

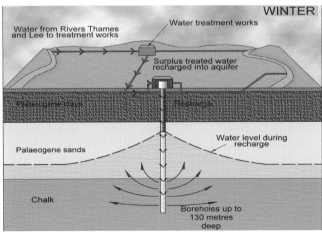

Figure 6.2 Lee Valley artificial recharge and recovery: summer injection phase; winter abstraction phase. Image credit: UK Groundwater Forum.

an extra resource of $10^5 m^3$ per day during the drier summer months. Similarly, in the Netherlands, water from the River Rhine is pumped onto lagoons in coastal dune systems, later to be recovered for public water supply. The managed groundwater mounds in the dunes also act as a flood defence system against extreme high sea tides.

Groundwater and river flow

Groundwater is used to augment flows in the River Clwyd in North Wales, in order to mitigate the impact from an adjacent public water supply abstraction scheme. This abstraction from the Sherwood Sandstone aquifer has minimal impact on the river during winter months, but during the summer increased demand for water, coupled with naturally lower river flows, may cause derogation. The **augmentation** scheme, the only one in Wales, was implemented by the Welsh Water Authority in 1977 at a cost of £580,000. It is designed to augment flow and ensure no reduction in the natural river flow downstream of the abstraction. This is achieved by releasing groundwater from several augmentation boreholes into the River Clwyd higher up in the catchment. These boreholes take advantage of the natural artesian conditions that exist in the area due to the thickness of superficial deposits with low permeability confining the sandstone bedrock aquifer. The scheme operates when river flow immediately upstream of the public abstraction drops below a trigger value of $147\,000\,m^3 d^{-1}$. At this point the abstraction must then be compensated by an equal augmentation discharge from the boreholes.

The River Darent in North Kent, in southeast England, is a good example of the interchange between surface and groundwater induced by groundwater abstraction. The river catchment lies almost wholly in the Swanscombe Chalk Block between the scarp of the North Downs and the Thames Estuary. It is bounded to the east by the River Medway and to the west by the Palaeogene cover of the central London Basin. The Swanscombe Chalk Block comprises a north-facing chalk dip slope that is partly concealed by Palaeogene in the northernmost part of the outcrop adjacent to the coast. The Chalk block is drained by the rivers Darent and Cray. Swanscombe is centrally situated within the Thames Gateway development area, has access to the high-speed rail link to the Continent, the motorway system and Thames port facilities, and is one of the fastest developing areas in the UK. Although demand from

water-intensive industry has declined in the last forty years, overall water demand is again increasing.

Given this background, the River Darent attained notoriety in the early 1990s when river flow ceased during dry summer months. The abstraction of groundwater for public supply from the Chalk aquifer along the Darent was believed to interfere with river flow, and the water companies were instructed by the regulator, the Environment Agency, to find alternative sources that did not derogate the surface water. Thames Water Utilities Limited, the water supply company, needed to replace sources amounting to $27\,Ml\,day^{-1}$ isolated from the river. Following a thorough investigation of the whole chalk block, this was achieved and the new boreholes are now in supply. The River Darent now has a perennial flow.

Rights and licensing

Norman Law has it that any water on or beneath a landowner's property is his by right – a laudable concept, but not a workable one, given the ever-increasing demand for fresh water for domestic, agricultural and industrial use. Nevertheless, this form of law prevailed in many areas until recently. For example, in the Channel Island of Jersey, Norman Law was only replaced by the new Water Resources (Jersey) Law in 2007. Thereafter, registration of all groundwater sources became mandatory and licensing of groundwater sources abstracting greater than $15\,m^3$ per day was required. This achieved two things for Jersey: firstly, it allowed the regulator some insight as to the volume of groundwater that the larger users were abstracting, and secondly, it gave the regulator some control over the larger abstractors. Licences of right were issued initially and these have since been moderated according to actual abstraction so that a licensed abstraction system is slowly evolving that does not allow withdrawal over the perceived safe yield of the aquifer. The aquifer is essentially weathered crystalline basement and borehole yields are, for the most part, modest. There are, however, some larger yielding boreholes and it is these that are now licensed. Registration of the smaller sources provides some additional abstraction information where, before the Water Resources Act, there was none.

Source licensing allows the regulatory body to apportion groundwater to users in an equitable manner. The regulator tries to ensure that the most productive uses of groundwater are prioritized, although not wholly at the expense of lesser users. This ensures that optimum usage is promoted and the aquifer itself is safeguarded from over-exploitation. In some countries licences are issued by the regulator and the licence holders are encouraged to trade their licences to the highest bidder, again a means of promoting optimum use of the resource.

River basin management issues
The Orange–Senqu River

The scale of a management area should not matter in itself. However, in many parts of the world large-scale river basin management is fraught with problems and issues. The Orange–Senqu River basin in southern Africa, for example, stretches almost the whole width of South Africa westwards to the sea in Namibia. It has an area of nearly one million square kilometres and includes a wide range of geological, geomorphological and environmental features as it passes from the Orange River headwaters in the Lesotho Highlands in the east towards the Namib Desert and the Atlantic Ocean in the west (Fig. 6.3). Long-term average rainfall over the basin ranges from as little as 200 mm in the arid lowlands in the west to over 1000 mm in the Lesotho Highlands. There is

considerable annual variation in rainfall with periodic cycles of drought years. Groundwater resources are threatened by demography, pollution, land use change and over-abstraction. The geological and hydrogeological setting for the basin includes two areas containing sediments of Karoo age and an otherwise irregular distribution of varied rock types ranging from the Archaean basement rocks to the Quaternary Kalahari Beds. The upper part of the Karoo sequence contains a significant water-bearing rock sequence, while the Transvaal dolomites are also an important aquifer. Other strata contain some groundwater and offer generally small yields to support rural needs and the mining industry.

Sustainability of supply is an issue, with ever-increasing demand set against periodic drought climate cycles. Estimates of recharge are complicated by the low effective rainfall in the arid and semi-arid areas, and by the variability of rainfall in the more humid parts of the basin. Long-term average values for recharge range from just 1 mm/year in the west to 100 mm/year in the east.

Overall understanding of the aquifer systems is patchy. Some areas have been studied in detail with comprehensive and validated numerical flow models, while in others

Figure 6.3 The Orange River on the border between South Africa and Namibia.
Image credit: Shutterstock/Vickyvv.

there is not even a conceptual understanding of flow. Water balance estimates are confused by difficulties in estimating recharge so that the overall capacity of most aquifers is largely unknown. Groundwater quality, pollution and groundwater protection are also issues in the Orange–Senqu River basin. **Acid mine water** discharges are a current problem in some parts of the basin. Elsewhere shallow groundwater systems are liable to pollution from poorly sited pit latrines and other foci of surface and near-surface pollutants (Fig. 6.4). **Disperse pollution** of groundwater from the application of nutrients and pesticides by the agricultural and forestry industries is also prevalent.

Data scarcity hinders detailed evaluation of the groundwater systems to be managed. Programmes of exploratory investigation, drilling and geophysics, will greatly help understanding of major and minor aquifers alike. However, the key to improved data and subsequent understanding is routine data gathering during drilling, testing and monitoring, supported by a user-friendly and accessible database.

The database needs to be shared by all stakeholders, regardless of international borders. The data gatherers and data users need to work to a common protocol and to a common and shared level of understanding so that a sensible dialogue can be maintained between them. This is essential if the overall surface and groundwater resources of the entire river basin are to be managed to develop an optimum sustainable yield. Drought and flood are problems that need to be coped with, and strategies to reduce demand in times of drought need to be put in place. Environmental protection from over-use of water resources is a vital component of the management strategy. **Transboundary aquifers** are not perceived as a priority issue at the moment.

For all these issues there are technical solutions. But technical solutions cost money, and the limiting factor is not 'can it be done' but 'can it be afforded'. Cost-effectiveness is, therefore, vital. Technical solutions include improved data gathering and data management, enhanced monitoring networks, sharing of data and understanding, establishment of groundwater protection zones, protection of the environment from acid mine drainage and many other do-able projects. Improved access to data will enable more sophisticated investigatory techniques to be applied, leading towards numerical modelling supported by historical data. Investigation of recharge is a priority, as without

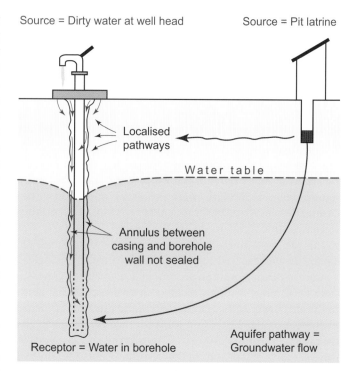

Figure 6.4 Contamination sources to village water supply borehole with pit latrine in the source capture zone.

this knowledge sensible management of the groundwater resource is not feasible.

The challenges that are to be faced over the sustainable management of the groundwater resources within the Orange–Senqu River basin are significant, but none are insurmountable. The management issues fall into two groups: physical measurement and monitoring to support understanding of the groundwater flow systems, and the basin-wide governance and oversight of the resources within the basin.

7 Groundwater quality

Naturally potable and health-giving

The slogan 'Water for life' is used frequently; the all-important concept is that of drinking water quality and health. Indeed, the period 2005 to 2015 was designated the United Nations 'Water for Life' International Decade for Action, designed to implement some of the so-called **Millennium Development Goals**. These strive to provide potable water supply and better sanitation coverage for the world's poor, a basic but essential objective.

Groundwater that is not contaminated by a man-made surface or near-surface contaminant will usually contain neither harmful **inorganic compounds** nor pathogenic bacteria and viruses. For the most part, groundwater discharging from springs or seepages and abstracted from wells and boreholes is safe to drink. This is largely because of the natural filtration processes the groundwater has been subject to during transport through the aquifer medium to the point of discharge or abstraction. The World Health Organization *Guidelines for drinking-water quality* (2011, p54) underlines the importance of groundwater as a drinking-water source:

> Groundwater from deep and confined aquifers is usually microbially safe and chemically stable in the absence of direct contamination. However, shallow or unconfined aquifers can be subject to contamination from discharges or seepages associated with agricultural practices (e.g. pathogens, nitrates and pesticides), on-site sanitation and sewerage (e.g. pathogens and nitrates) and industrial wastes.

However, as a result of reaction during the transport of groundwater through an aquifer, the groundwater may naturally build up solutes that can, in some instances, exceed World Health Organization drinking-water guideline values. Concentrations above 0.01 mg arsenic l^{-1} and 1.5 mg fluoride l^{-1} are, for example, cause for concern in many parts of the world, especially parts of South Asia, Africa and South America (Figs 7.1; 7.2). High iron and manganese concentrations may be

Figure 7.1
Bangladesh: arsenic in groundwater can have serious effects on health if exposed over a prolonged period. Image credit: Jeff Davies.

unacceptable for some uses such as laundry and some industrial processes, but they are not a danger to health.

Element deficiencies in drinking water, as well as excesses, can be detrimental to health. Dietary iodine deficiency, for example, can cause goitre and cretinism. Drinking water alone is unlikely to be the cause of iodine-deficiency disorders, but it can contribute to an overall deficient dietary intake. Iodine-deficiency disorders have also been linked with lack of selenium. Deficiency in fluoride can be associated with dental caries. Indeed, in some places this has resulted in the use of artificial addition of fluoride to the public water supply to promote dental health. Yet for many if not all chemicals, excesses can be equally detrimental. At fluoride concentrations much above 1.5 mg Fl^{-1}, long-term exposure through drinking water can cause fluorosis; the window for optimal concentrations of fluoride in drinking water is, therefore, relatively narrow.

58

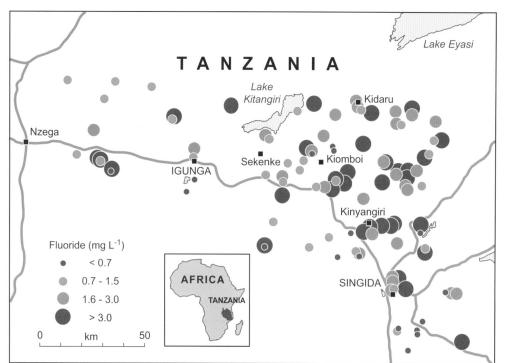

Mineralization

The mineralization of pristine groundwater is controlled by a variety of biogeochemical processes so that a broad range of groundwater types and ranges of concentration occur (Fig. 7.3). Generally, 99% of the chemical make-up of uncontaminated groundwater comprises nine parameters, the **major ions**: the anions HCO_3, Na, Ca, SO_4, and cations Cl, NO_3, Mg, K and Si. The negatively charged anions will balance the positively charged cations when expressed as **millequivalents** or equivalents per million. These can be plotted graphically on a trilinear **Piper Diagram** in order to compare sample clusters and sample trends: for example, between groundwater and seawater for a coastal aquifer (Fig.7.4).

For the most part, however, concentrations are expressed simply as parts per million (by weight) which is the same unit as mg per litre for weakly mineralized water. The major groundwater type is usually dominated by $Ca\text{-}HCO_3$ (calcium bicarbonate). As direct rainfall recharge percolates underground it moves downwards under gravity towards the water table. Rainfall is naturally acid although weakly buffered, i.e. the pH can easily be raised, and the acid attacks

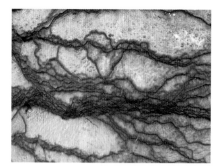

Figure 7.3 Tufa deposit with red iron bacteria lobes at a limestone spring in South Wales. Image credit: Jeff Davies.

the minerals present in the rock. The most reactive minerals, including particularly the carbonate mineral calcite, have the greatest impact on water chemistry, taking firstly Ca (calcium) and HCO_3 (bicarbonate) into solution and buffering the acidity so that the pH is less easily modified.

Residence times

Residence times underground vary from a few months to tens of thousands of years and more (Fig. 7.5). Groundwater

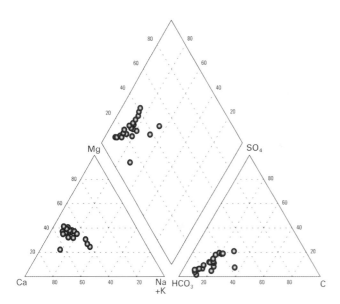

Figure 7.4 Piper diagram showing a cluster of Ca-HCO₃-dominated groundwaters from a single aquifer unit.

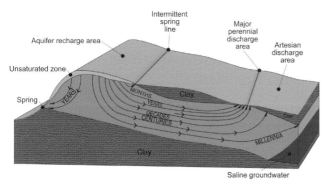

Figure 7.5 Different flow paths reflect different ages of groundwater or residence times within the aquifer. Image credit: UK Groundwater Forum.

attains a pH that reflects the containing rock mass. This is most commonly buffered at around pH 6 to 8 but extremes can occur. Acid buffering is less well achieved in crystalline, non-carbonate bedrock conditions. Older waters, be they acid or alkaline, eventually acquire a chemical equilibrium with the containing rock mass; compositions evolve by dissolution, precipitation and cation exchange, such that Ca-HCO₃

(calcium-bicarbonate) waters can evolve into Na-HCO₃ (sodium-bicarbonate) types and ultimately to Na-Cl types. Increasing residence time leads to additional build-up of numerous trace elements including alkali and alkaline earth metals.

Redox reactions

Another significant property of groundwater arises once it loses direct contact with the atmosphere and is transported deeper into the aquifer. The **redox** state of the water is controlled by the quantity of oxygen and other oxidizing compounds present in the system. Reduction of these by chemical and microbial reactions can result in an anoxic groundwater with significant chemical differences from its oxic counterparts at and near the ground surface. Sequential changes take place down the groundwater flow gradient following loss of dissolved oxygen (O₂), which can see first the loss of NO³ (nitrate) and thereafter the introduction of manganese and iron. Extremes in redox reaction progress can ultimately see the loss of SO₄ (sulphate) and the introduction of NH₄ (ammonium). The redox environment of the groundwater also plays a key role in determining the abundance of **trace elements** such as arsenic, chromium, selenium and uranium, many of which affect the suitability of water for drinking.

Conductivity

The ability of mineralized water to conduct electricity is described in terms of its **specific electrical conductivity (SEC)**. This is a function of temperature, the type of ions that are present and the total concentration of ions. Value adjusted to 25°C reduces the dependence of SEC to just type and abundance of ions. At this temperature pure water has a conductance of 0.055 micromhos per cm and groundwater will usually lie in the range 30 to 2 000 micromhos per cm. Ca-HO₃ and Ca-SO₄ type groundwaters generally have moderate conductance whereas Na-Cl type waters have a higher conductance for a given total concentration of ions. Specific electrical conductivity is a valuable indicator of total concentration of ions: SEC in micromhos per cm multiplied by 17 approximates to total concentration of ions in mg/l.

pH values

The hydrogen ion concentration is expressed as pH. A neutral water has a pH of 7, reflecting an equal number of H⁺ and

Figure 7.6 Measuring the pH of an acid minewater discharge, Gauteng, South Africa. Image credit: Jude Cobbing.

OH⁻ ions; an alkali water has a pH >7 reflecting a preponderance of OH⁻ ions; and an acid water has a pH <7 and a preponderance of H⁺ ions. Most groundwater lies within the pH range 5 to 8. High values are usually associated with $Na-CO_3-HCO_3$ type waters, whereas moderately high pH values are normally associated with HCO_3 dominated waters. Low pH values (<4) are normally associated with acid groundwaters (Fig.7.6).

Groundwater age

Groundwater age dating uses the known decay rates of radioactive isotopes, the timing of the introduction into the atmosphere of isotopes from nuclear testing or reactors, or the history of the release of manufactured gases to estimate the age of a groundwater sample. The water molecules in the sample carry no age information themselves, so a concentration of a marker, a parent or daughter isotope, or a manufactured gas serves as a proxy for age. In common practice, each dating method is formulated assuming the sample has not been in contact with the atmosphere since it fell as rainfall to infiltrate the ground surface. When collecting samples, these should not be in contact with the atmosphere. Each dating method provides a value or range of values for sample age despite the likelihood that the sample is a mixture of waters, and the age is an aggregate of all these, the components of

which are of different ages, having travelled along different flow lines.

Recharging groundwater contains radionuclides from the atmosphere, which, once the water infiltrates the subsurface, decay asymptotically to zero concentration. Of the methods based on asymptotic decay, the best-known is the radiocarbon, or ¹⁴C, method, useful for dating groundwater less than approximately 50 000 years old. Cosmic rays produce ¹⁴C naturally in the atmosphere and the isotope, which has a half-life of 5730 years, dissolves as CO_2 in rainfall and in moisture in the root zone. The ³⁶Cl method is in many ways similar to the radiocarbon technique, although the isotope's half-life of approximately 301 000 years is considerably longer than for ¹⁴C, and hence the method is appropriate for dating much older groundwater. Cosmic rays also produce small amounts of ³⁶Cl in the atmosphere, which dissolve in rainfall as chloride and decay slowly in water flowing through the subsurface. This provides yet another age-dating method.

Historical matching techniques use radionuclides and anthropogenic chemical compounds that were rare or non-existent in pre-industrial times and have been released recently to the atmosphere, where they circulate, dissolve in rainfall, and infiltrate the subsurface. Some of the isotopes and compounds are sufficiently unreactive and trace the transport of water well enough to serve as proxies for age. These markers have appeared within about the past 50 years and are used to date young groundwater. The concentration of each marker has a unique history in the atmosphere, and in meteoric precipitation. Some markers appeared in pulses marking specific events; others have accumulated steadily with time. Using the **piston flow** model, interpreting groundwater age from marker concentration is straightforward: the concentration is compared to the known history to find one or more recharge dates that match. Nuclear weapons testing between the 1950s and mid-1960s released radionuclides to the atmosphere and are notable because they form a 'bomb pulse', a period during which groundwater recharge was rich in isotopes. The most persistent is tritium (³H). After atmospheric testing was banned, the isotopes diminished in concentration as they entered the oceans and subsurface.

Groundwater enriched in the bomb isotopes could only have recharged an aquifer during the bomb pulse, and hence is known to have an age falling within a set range. Assigning a more specific age to a sample can be complicated, however, because an observed concentration matches two points in time, one on the rising and one on the falling limb of the pulse. A more distinctive isotope is krypton (^{85}Kr), with a half-life of 10.76 years, which is released during the operation of nuclear reactors. Its concentration in the atmosphere has increased in a nearly linear fashion over the past several decades. Accordingly, when radioactive decay is taken into account, a specific ^{85}Kr concentration measured in groundwater will match a unique age.

Among the manufactured gases, chlorofluorocarbon compounds, or CFCs, and sulphur hexafluoride (SF_6) have proved useful for age dating. The various CFC compounds have different histories. Some increased in concentration early and then levelled off as they were banned from use, while others have accumulated more recently. The atmospheric concentration of SF_6, used primarily in electrical equipment and the manufacturing of semiconductors, has increased steadily since the mid-1960s. The concentration in groundwater of any one of the manufactured compounds thus generally maps to a unique sample age.

A key issue with age dating is that the signature of recharging water may be contaminated at a later date by more modern isotope concentrations. As a consequence, these techniques tend to give younger dates than you would get from a Darcian groundwater flow model.

Groundwater provenance

Most groundwater is recharged by the direct infiltration of precipitation and of surface water, rivers and lakes, or by subsurface inflow, and thus originates from precipitation. Owing to evaporation and exchange processes, the isotope content of the original rainfall can change during the transition from precipitation through the soil to the unsaturated or vadose zone. The average precipitation isotope data and the distribution of the stable isotopes within the groundwater isotope composition help to identify the groundwater flow system; stable and radioactive environmental isotopes are commonly used to support groundwater investigations. Environmental isotopes include the heavy isotopes of the elements of the water molecule, hydrogen, (^{2}H) deuterium and (^{18}O) oxygen occurring in water as constituents of dissolved inorganic and **organic compounds**. A molecule of water normally consists of two atoms of hydrogen and one atom of oxygen. Most of the hydrogen atoms have atomic

mass one (H), but a small number occur as deuterium and the normal ^{16}O oxygen is sometimes replaced by ^{18}O. Applications of stable isotope ratios of hydrogen and oxygen in groundwater are based primarily upon isotopic variations in atmospheric precipitation, as recharge to a hydrogeological system.

The isotopic composition of a water sample is expressed as the per mille deviation of the isotopic ratio, deuterium/hydrogen or ^{18}O/ ^{16}O from that of a standard. The standard of reference in general use is an arbitrary point of reference called **SMOW (Standard Mean Ocean Water)**. Whenever water condenses or evaporates, an isotopic fractionation occurs, mainly because the heavy isotopic components, deuterium and ^{18}O, have lower vapour pressures than H and ^{16}O. This deviation readily identifies those groundwaters derived from lakes and rivers that have been standing and subject to evaporation from those that derive from direct rainfall recharge. They also identify seasonal changes in rainfall recharge type with depleted summer rainfall and normal winter rainfall. This enables samples to be collected from the unsaturated zone to identify annual recharge patterns and rates of infiltration, i.e. to determine actual recharge values.

Organic chemistry

A large variety of organic compounds can be found where groundwater is contaminated by industry or by intensive agriculture. Some of these compounds may be harmful to health if consumed, and need to be eliminated from sources used for domestic and animal consumption. They may include volatile organic compounds used in industry, herbicides, insecticides, fungicides, rodenticides, and algicides, also plasticizers, chlorinated solvents, benzo-pyrene, and dioxin (Fig. 7.7). Many of these compounds break down in groundwater to form so-called daughter products. The original compound and its likely daughters need to be analysed to determine their presence in the sample of groundwater under investigation. This can be expensive, so it is useful to determine first of all which products are likely to arise in groundwater in a given area. This depends on the types of industry and farming that take place locally, but it allows a likely suite of compounds and daughters to be determined. Specialist techniques are required to sample for organic compound determination.

Indicators of organic contamination include elevated nitrate, sodium and sulphate concentrations. These surrogates

Figure 7.7 Pesticide spray onto a wheat field. Image credit: Shutterstock/PolakPhoto.

can be used to identify areas of contamination that need to be surveyed. Organic pollutants cover a wide field of compounds; in some instances trace concentrations can be damaging to health if consumed on a regular basis. From time to time products are banned from use, but may still occur dissolved in groundwater many years after their application has ceased.

Bacteria

Coliform bacteria occur naturally in the environment from soils and plants and in the intestines of humans and other warm-blooded animals. They can be used as an indicator for the presence of pathogenic bacteria, viruses, and parasites from domestic sewage, animal waste, or plant or soil material. Again, analyses of these forms of contaminant can be expensive and, as with the organic contaminants, require specialist sampling techniques.

Groundwater sampling and analysis

The purpose of groundwater sampling is to collect representative samples of the aquifer at the well location. When selecting the sampling approach to be used, the investigator should consider characteristics of the contaminant(s) such as volatility, solubility, density (denser or lighter than groundwater) and their resultant fate in the subsurface (adhesion to soil particles, biodegradation, etc.). First of all, the sample

collected must be representative of the environment being investigated; this requires prior knowledge of the prevailing groundwater flow system and of the abstraction source. Many **determinands** require that the sample is not contaminated by contact with the atmosphere or that it is allowed to de-gas. Others require that the sample is filtered, usually through a 45 µm filter, or that it is acidified to ensure stability before the sample reaches the analytical laboratory. It is also usual to keep samples refrigerated between collection and laboratory analysis to discourage chemical change and bacterial activity.

Some determinands need to be analysed at the well-head. It is usual to pass the sample through a flow-through cell containing electrodes to measure dissolved oxygen (O_2) redox potential and pH, as these are all susceptible to change on contact with the atmosphere. Bicarbonate (HCO_3) needs to be determined on site by titration, as it cannot otherwise be stabilized. Temperature must also be measured in situ before the sample has had time to equilibrate with the temperature of the air. Other determinands can be measured in batches in the laboratory to provide the list of analyses required. Before collecting samples for analysis, contact the laboratory and ask for instructions regarding containers, on-site processing and handling – sampling protocol does vary from laboratory to laboratory, and different protocols exist for different analyses. For example, samples for CFC determination must be collected in a glass-stoppered glass jar and sealed in a tin containing the same sample with air totally excluded from the containers.

The laboratory will provide a description of the analytical methodologies that are used and will provide evidence of their own competence as judged by an external regulatory authority.

Guernsey – a case study

Guernsey is a small island in the British Channel Islands off the Atlantic coast of Northern France. It is roughly triangular in shape and is 10 x 10 x 14 km and rises to some 107 m above sea level. The groundwater body on the island consists of three layers (Fig. 7.8). Where present, there is an upper granular aquifer within superficial deposits of alluvium and raised beach material. Beneath this is the main aquifer, which is contained within the shallow weathered zone of ancient crystalline metamorphic rocks. Beneath this is a deeper aquifer with groundwater restricted to occasional dilated fissures but without any significant primary porosity or significant storage potential.

Twenty-one pumped groundwater samples were collected and submitted for mineralogical analysis; four samples were taken for CFC analysis, and a further ten for stable isotope investigation. The data show that the groundwater contains

a small element of salinity derived from sea spray and dry deposition, but there is little evidence of actual marine invasion of the aquifer. The groundwater samples were otherwise moderately mineralized, oxygenated, with specific electrical conductance in the range 427 to 1578 µS/cm. Na and Cl were elevated above the trend between local rainwater chemistry and seawater, while Ca and Mg were also elevated due to water/rock interaction during the maturation of the groundwater. Over half of the samples contained nitrate (NO_3-N) at concentrations greater than the European Community maximum admissible concentration for drinking water. Some of the nitrate may derive from leaking cesspits, but application of nitrogen fertilizer to cultivated land almost certainly accounts for the remainder. The average pH of the samples was on the acid side of normal at 6.6, reflecting the lack of carbonate material available in the aquifer, other than shelly debris within parts of the superficial cover.

Attempts at groundwater dating by analysis of CFC species were hindered by local contamination, particularly near the airport. As the pumped samples are likely to represent mixed waters of different provenance, interpretation is difficult. However, 75 to 100% of the water was identified as young, recently recharged water, in keeping with shallow and short flow paths within the main aquifer. One sample indicated between 40 and 50% of the sample being of the young

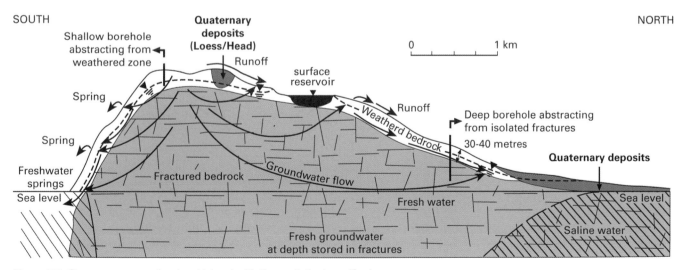

Figure 7.8 Guernsey: a small upland island with three distinct aquifer layers.

component, despite there being no indication of any older groundwaters or deeper groundwater circulation.

The stable isotope data for all but one sample ranged between -5.94 and -5.18 per mil with respect to Standard Mean Ocean Water (SMOW) for ^{18}O and between -35.9 and -31.4 for ^2H deuterium. These indicate an element of evaporation prior to recharge. One exceptional borehole sample had a value of -3.19 per mil with respect to Standard Mean Ocean Water for ^{18}O and -24.3 for ^2H deuterium. This sample was taken from a shallow borehole near to a small lake. It is likely that significant evaporation takes place from the lake, especially during the summer months, and that much of this water percolates into the aquifer to source this one borehole.

The outcome from the study was that much of the groundwater on the island was contaminated as a result of intense horticultural and agricultural activity, as well as leaking cesspits. In addition, the groundwater chemistry has a strong maritime influence although physical seawater intrusion is not apparent. This derives principally from sea spray. Stable isotope data do not contradict the idea that slight variations in groundwater chloride concentration reflect the influence of seawater. However, there is evidence that evaporated surface water has also entered the aquifer, albeit only seen at one site near a lake. The high levels of CFC species near the airport are indicative of organic pollution by materials used on site including degreasants, jet fuel, fire-fighting foam and other chemical substances.

The hydrochemical data were used to support the conceptual groundwater flow of the island. The total project cost (in 2007) was about £60,000. The study was used to underpin the management of the island resource in order to protect the various public supply sources that are in use throughout the year.

Spas, bottled waters and deep groundwater circulation

Although the pressure of overlying rock tends to reduce the numbers of open fractures with depth, there may be a modest volume of deep groundwater circulation along selected flow-paths. The emergences of such deep flow paths often represent mineral and thermal water springs, some of which were developed in the past into health spas or tourist attractions (Fig. 7.9). Deep flow-paths may occur in an area of tectonic disturbance such as a major fault, but also require sufficient head to drive water down into the earth. It can take many years for the water to re-emerge from a deep circulatory system; radiometric dating estimates of the groundwater rising at Buxton in the English Peak District, for example, suggest an age of 10 000 years. Clearly, such old water should be free from modern day contaminants, and is likely to be in mature hydrochemical equilibrium with the rocks through which it has passed. It could, therefore, be quite saline, but happily the Buxton water is only modestly mineralized and is widely enjoyed as a bottled natural mineral water that predates the atmospheric pollution of the post-Industrial Revolution era.

In Central Wales, the spa resorts of Builth Wells, Llandrindod Wells and the lesser known Llanwrtyd Wells and Llangammarch Wells all relied on old upwelling groundwater from Silurian/Ordovician rocks, driven by the head provided by the surrounding hills. The sources are of variable hydrochemical type, from saline to iron-rich (chalybeate) and sulphur-rich. These small spring discharges relate geologically to the Tywi Lineament, with groundwater circulating down to 300 m before rising to mix with shallower waters. Mixing with the shallower groundwaters tends to disguise the chemistry of the deeper circulating waters. Known to the Romans for their curative powers, the spa waters were drunk warm and by the pint by the Victorians, and are now a novelty on display for visitors. Treatments available at the spas also included the needle shower (high-pressure needles of saline water jetted at the naked patient) and other rather odd Victorian remedies. Similar deep-seated saline groundwater systems occur in British basement rocks in the Lake District, at Wentnor near the Shropshire Long Mynd, and at a number of sites in Scotland.

Other than that at Buxton, few British bottled waters derive from old, deep groundwater circulation. Most come from relatively shallow sources. For example, waters bottled as Natural Mineral Waters (according to EC labelling requirements) in Scotland issue from springs or are pumped from boreholes in Devonian and Carboniferous sandstone and lavas, one (Caithness Spring, Berriedale) from the Pre-cambrian and one (St Ronan's Spring, Innerleithen) from a borehole in Silurian shales. Another source in lavas is extremely weakly mineralized and represents a very young water from a short flow-path.

Figure 7.9 The Cave and Basin at Banff, Alberta, is one of nine sulphurous hot springs located along the Sulphur Mountain thrust fault in Devonian limestone. The water circulates to an estimated depth of 3 km. Image credit: Fiona Robins.

8 Groundwater pollution, vulnerability and protection

Types of groundwater degradation

Degradation of groundwater is the result of anthropogenic intervention, which can be:

- Over-exploitation where groundwater levels become unacceptably low. This not only reduces borehole yields but may permanently damage the aquifer. It may also cause ingress of saline water from the sea or from upwelling from deeper saline aquifers.
- Inappropriate activities such as point sources of uncontrolled waste disposal and spillages, or diffuse sources such as application of pesticides or fertilizer to crops (Fig. 8.1), or even nuclear fallout.
- Land use changes such as deforestation, deep ploughing or urbanization.

By far the largest cause of degradation is pollution caused by agriculture, industry and urbanization (Fig. 8.2). In most cases, contamination of groundwater takes place almost imperceptibly. Slow movement of water from the surface through the unsaturated or vadose zone to the water table means that it may be many years after a chemical or fuel spill

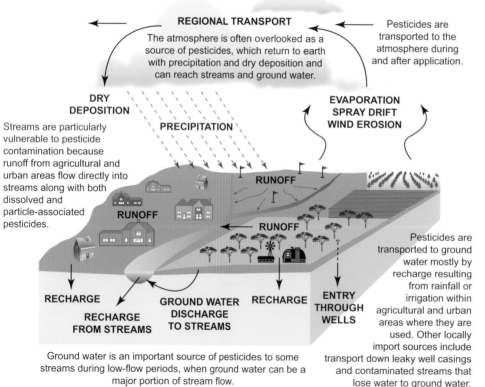

REGIONAL TRANSPORT
The atmosphere is often overlooked as a source of pesticides, which return to earth with precipitation and dry deposition and can reach streams and ground water.

Pesticides are transported to the atmosphere during and after application.

DRY DEPOSITION

EVAPORATION SPRAY DRIFT WIND EROSION

PRECIPITATION

Streams are particularly vulnerable to pesticide contamination because runoff from agricultural and urban areas flow directly into streams along with both dissolved and particle-associated pesticides.

RUNOFF

RUNOFF

RUNOFF

RECHARGE

RECHARGE FROM STREAMS

GROUND WATER DISCHARGE TO STREAMS

RECHARGE

ENTRY THROUGH WELLS

Pesticides are transported to ground water mostly by recharge resulting from rainfall or irrigation within agricultural and urban areas where they are used. Other locally import sources include transport down leaky well casings and contaminated streams that lose water to ground water.

Ground water is an important source of pesticides to some streams during low-flow periods, when ground water can be a major portion of stream flow.

Figure 8.1 Pesticides arrive in surface and groundwater mainly through runoff and direct rainfall recharge. Image credit: United States Geological Survey.

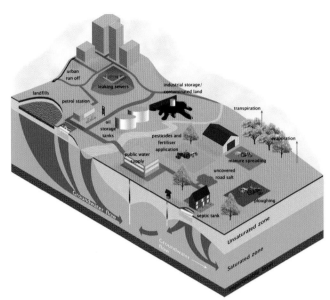

Figure 8.2 Typical pathways for surface contaminant hazards to arrive in groundwater. Image credit: UK Groundwater Forum.

occurs at the surface before it is recognized in a groundwater supply. Past application of nitrate fertilizer to land may cause persistent elevated levels of nitrate in groundwater long after application has ceased or been moderated. Microbial contamination indicates a faster travel time, as many pathogenic organisms have limited persistence.

Some **attenuation** of contaminants will occur as a pollutant moves down through the soil and vadose zones (Table 8.1). The soil zone is important because of the abundant bacterial activity in soil, which helps to break down some pollutants,

and the presence of carbon, which can inhibit the transport of many pollutants by **sorption**.

Attenuation of microbial contaminants is largely dependent on their persistence. Some viruses can survive in groundwater for up to 150 days. Microbial contaminants may be removed by filtration if they are too large to migrate through pore spaces in the rock or by adsorption to clay minerals. Attenuation of inorganic chemicals, such as heavy metals, can occur through precipitation with the formation of carbonates, sulphides or hydroxides. Degradation of many organic compounds can occur through sorption and **biodegradation**. Light aromatic organic substances such as petrol and diesel, known as **LNAPLs**, will float on top of an aquifer, whereas heavy compounds such as the many **aliphatic compounds** used as solvents, known as **DNAPLs**, will sink to the bottom of the aquifer.

Groundwater vulnerability to pollution

Groundwater pollution risk is the interaction between the contaminant load applied to the aquifer and the natural pollution vulnerability of the aquifer. Groundwater vulnerability is based on the potential attenuation capacity between the ground surface (or point of pollution, which may be underground beneath the soil layer in the case of leaking cesspits and pit latrines) and the water table. Obviously confined aquifers are least vulnerable and can only be contaminated by transport from the unconfined recharge zone to the confined part of the same aquifer system. Four classes of vulnerability are recognized in the United Kingdom:

- Extreme: Vulnerable to most water pollutants with rapid impact by many polluting activities (associated with fractured aquifers with a shallow water table).

Table 8.1: Process causing attenuation of pollutants underground

		Soil zone	Unsaturated or vadose zone	Saturated zone
Dilution		Minor	Minor	Major
Retardation processes	*Sorption*	Major	Minor-significant	Minor-significant
	Ion exchange	Significant	Significant	Significant-major
	Filtration	Major	Significant	Significant
	Precipitation	Minor-significant	Significant	Minor-significant
Elimination processes	***Hydrolysis***	Significant-major	Significant	Significant
	Volatilization	Major	Minor	Minor
	Biodegradation	Major	Minor-significant	Minor-major

67

- High: Vulnerable to many pollutants except those highly adsorbed or readily transformed.
- Low: Only vulnerable to the most persistent pollutants over a long time period.
- Negligible: Confining beds present with no significant flow from shallower aquifers.

An aquifer vulnerability map is a useful planning tool and shows areas of each of the four degrees of vulnerability. The maps are derived from the aquifer characteristics such as the depth to the water table, permeability and degree of fracturing.

Groundwater protection from agricultural activities

Groundwater vulnerability zonation is a valuable management tool designed to inhibit the application of nutrients to the ground. The 1991 European Commission Directive on nitrate pollution originally resulted in the designation of 72 **nitrate vulnerable zones (NVZ)** in catchments in England and Wales where nitrate levels exceed 11.3 mg-N/l or were likely to do so in the future. These covered 650 000 ha, mostly in central and eastern England. In December 2000, the European Court of Justice ruled that the UK had failed to implement the Nitrates Directive fully, stating that the Directive applied to all ground and surface fresh waters, so as to reduce the risk of eutrophication as well as to protect drinking-water sources. The UK Government, including the Welsh Assembly, were forced to take action to comply with the Court's judgement. This meant that the UK needed to complete the implementation of the Directive fully, or face the prospect of very substantial daily non-compliance fines. Government bodies began consultations on their proposals to extend the area of NVZs. Under the new rules additional NVZs covered about 47% of England.

Farmers in the NVZs are required to limit their applications of organic manure. The limit is 210 kg of total nitrogen per hectare for arable land, and 250 kg for grassland. After four years the limit must be reduced to 170 kg on arable land unless larger applications can be justified. The obligations to the farmer also include:

Do not apply inorganic nitrogen fertilizer between 1 September and 1 February unless there is a specific crop requirement. Do not apply it when the soil is waterlogged, flooded, frozen hard or snow covered. Do not apply to steeply sloping fields. Do not exceed crop requirement for the quantity of nitrogen fertilizer on each field in each year, having taken into account crop uptake, soil residue and organic manures. Do not apply fertilizers in a way that would enable them to directly enter surface water.

Applications of organic manures are not to exceed 250 kg/ha of total nitrogen over grass and 210 kg/ha averaged over the area of the farm not in grass each year. On sandy or shallow soils there are to be no applications of slurry, poultry manure or liquid digested sludge on grass between 1 September and 1 November and to fields not in grass between 1 August and 1 November. These manures should also not be applied when the soil is waterlogged, flooded, frozen hard or snow covered, or to fields that are steeply sloping or land within 10 m of surface water.

Similar zonation programmes have been put in place in many countries around the world. The objective is the same – to protect groundwater and surface water bodies from excessive nitrogen input. However, the details vary according to the prevailing climate and the types of crops that are commonly grown.

Protection of water from pesticide pollution is normally carried out through a set of guidelines detailing appropriate applications of particular compounds suitable for each crop type grown in a given country or state. Disposal of waste liquid pesticide and 'empty' containers is usually strictly controlled.

Groundwater protection from other activities

Polluting activities are widespread, and although rigid guidelines are in force in most countries, accidents occur and short-cuts taken that cause spillages and leaks. One of the more common forms of point source groundwater pollution derives from fuel storage, whether domestic oil fuel, a highway filling station or aviation fuel tanks at the airport. For example, at Heathrow airport near London, a major clean-up of the highly permeable Thames gravels had to be undertaken in the 1980s to recover several tens of thousands of cubic metres of Jet A1 fuel that had leaked from the storage and distribution system. On a similar but smaller scale, a petrol filling station at Bath in southwest England had a persistent leak over several years from a petrol storage tank. The leak was at a slow rate such that the deficit of fuel arriving and fuel leaving the station was written off over the years. The subsequent clean-up operation, however, was a major challenge as the fuel lay on top of groundwater in a sandstone aquifer underlying an intensely urbanized part of the city. These experiences led to a tightening up of the engineering of storage vessels, with all filling station tanks now being double lined. Domestic oil fuel storage remains an issue in most countries, where the onus is on the householder to ensure that his fuel system is tight and secure.

Other common hazards are roads and railways. Both use soakaways to disperse surface water underground, and these are ideal conductors of spillage from accident directly to a shallow groundwater system. Roads in icy conditions have salt applied to them; the salt goes into solution and again has access to groundwater via soakaways and nearby surface water courses. Railway ballast is regularly sprayed with herbicide to prevent weeds accumulating and disturbing the integrity of the track. Again, the application of pesticide can be limited according to the vulnerability zonation of the area. For example, a major public water supply wellfield operated by Thames Water at Gatehampton, some 30 km west of London, is adjacent to a mainline railway (Fig. 8.3). The railway authorities are not allowed to spray any herbicides 1 km either side

of the wellfield. Gatehampton is the largest groundwater abstraction scheme in Europe and draws from seven production boreholes drilled into the Chalk.

Industry is heavily regulated, as is the mining industry. Both activities are under strict obligations to prevent pollution to water bodies, but from time to time both sectors of industry suffer problems that do contaminate water bodies.

The problem with all this regulation is that it is only in force and properly policed in the richer countries. Many poorer countries are not adequately resourced to enforce their own regulations and guidelines, and this situation is exploited by some. This is a major issue that can only be resolved with international assistance to help those countries protect their surface and groundwaters from unscrupulous operators.

Groundwater source protection

As well as resource protection by vulnerability mapping there is also source protection that inhibits certain activities within the capture zone of important groundwater sources such as a public supply well or a bottled water source. For example,

Figure 8.3 Borehole No. 5 in the Gatehampton wellfield, Berkshire, England.

there are some 2000 source protection zones designated in England and Wales. Generally, there are three concentric zones of protection that can be applied around a source for a given and protected yield. In England and Wales they are defined as:

- Inner Protection Zone – the 50 day travel time from any point below the water table to the source. This zone has a minimum radius of 50 m. 50 days is selected as the time after which few if any viruses or microbial contaminants would survive underground to contaminate the source water, and many rapidly degrading organic pollutants would also be largely excluded from the source.
- Outer Protection Zone – a 400 day travel time from a point below the water table. This zone has a minimum radius of 250 or 500 metres around the source, depending on the size of the abstraction. This is designed to encompass a 400 day travel time to the source in order to exclude slow degradation of most pollutants.
- Source Catchment Protection Zone – the area around a source within which all groundwater recharge is presumed to be discharged at the source. In confined aquifers, the source catchment may be displaced some distance from the source. For heavily exploited aquifers, the source catchment protection zone is the whole aquifer recharge area where the ratio of groundwater abstraction to aquifer recharge (average recharge multiplied by outcrop area) is >0.75. However, there may still be a need to define individual source protection areas to assist operators in catchment management.

Other countries apply different rules to the same effect. In the Republic of Ireland, for example, the inner zone is defined as 100 days or 300 m, whereas in Italy it is 180 to 365 days depending on aquifer vulnerability and contaminant load, and in Australia it is simply 50 m radius. In all cases the zones are either determined manually, or with sufficient data a modelling approach may be applied.

The shape of the zones depends on a variety of factors:

- Aquifer thickness determines the transmissivity of the aquifer and the volume of water in the aquifer, and hence directly affects the area of 50 and 400 day zones and the shape of hydraulic capture zones. A 50% decrease in aquifer thickness results in an approximate doubling of the zone area and an increase in the width of the zone.
- Aquifer porosity has a direct impact on the area of the 50 and 400 day zones. A 50% decrease in porosity results in a 2–4-fold increase in the zone area.
- The hydraulic conductivity affects the shape of protection zones in terms of the width and the downgradient extent of the zone. An increase in the hydraulic conductivity will decrease the width of the capture zone to a borehole.
- The hydraulic gradient affects the width and downgradient extent of the hydraulic catchment zone – the steeper the gradient the narrower the zone.
- The abstraction rate directly affects the area of hydraulic protection zones. Interference between abstraction boreholes can greatly affect the shape of zones, producing 'tails' and 'holes'.
- The rate of groundwater recharge directly affects the area of the catchment zone. The areas of 50 and 400 day zones are less sensitive to recharge rates. Recharge from surface runoff from adjacent drift or karstic areas can distort the catchment zone.
- No flow boundaries such as faults and groundwater divides constrain the shape of hydraulic protection zones.
- Head-dependent boundaries affect the shape and possibly reduce the area of hydraulic capture zones (particularly the catchment zone).

Borehole and spring protection

Boreholes, particularly those situated near polluting activities, for example, a borehole drilled in a farmyard, need to be constructed in a way that prevents ingress of surface water. It is usual to place a sanitary seal around the uppermost section of the borehole casing to ensure a solid and watertight connection between the borehole casing and the ground. In addition, it is usual to set the part of the casing that emerges above ground into a concrete plinth. A primary consideration is the depth to which the sanitary seal must extend below surface. Most of the technical literature agrees on a minimum depth of 5 m, but all qualify this on the basis of the nature of the near-surface material and the proximity of compromising activities. The devil in the detail of this aspect is the requirement that the annular space be sufficiently large and open to receive the grout mixture without restriction for the entire length of the sanitary seal. In addition, the sanitary seal should rest on some form of underlying annular fill material, perhaps separated by tamping down canvas on the top of the fill material. In the case of open borehole completions, this may be bedrock into which the surface casing is embedded.

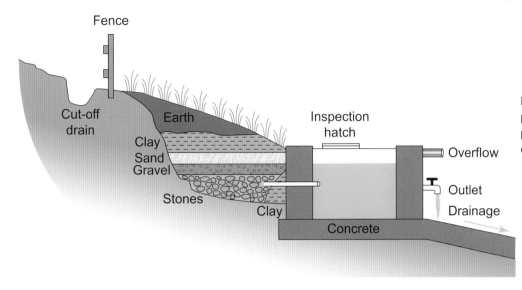

Figure 8.4 Typical spring protection box with feed pipe to consumer. Image credit: Jeff Davies.

Rural village hand-pump supplies in Africa and Asia are a special issue. These need sophisticated construction at surface to include drainage away from the source and prevention of access by animals. Similarly, spring sources need to be fenced off and the source 'eye' protected by boxing it in and backfilling with gravel with a protective clay layer above. A pipe let into the box and open to the gravel allows a protected capture of the spring water, which can then be gravity fed downhill to where it is needed (Fig. 8.4).

A borehole that is no longer in use (abandoned or decommissioned) and has been left open represents an unacceptable risk and potentially lethal threat to its subsurface environment. It provides an excellent pathway for pollutants to enter the subsurface and contaminate the groundwater resource. Old boreholes should, therefore, be capped off, while those that might allow transport of groundwater between a lower and an upper aquifer due to the head difference in the aquifers need to be backfilled with cement.

Groundwater contaminating the surface environment

There are occasions when contaminated groundwater discharges to surface and damages aquatic systems. The most widespread example of this is acid mine drainage. Similarly, salinization, which is the mineralization of undrained soils under irrigation, can destroy the soil structure through alkaline build-up.

Pollution from mining

Mining for coal and metals often continues below the water table, with water pumped out of the mine to surface. This water is usually reasonably uncontaminated groundwater. When a mine is abandoned, pumping ceases and groundwater rebound occurs, slowly flooding the mine workings. These environments are rich in pyrites and other metal sulphides. After the sulphides have been dewatered they oxidize to form soluble hydrous ferrous compounds. On mine abandonment the hydrous compounds dissolve as the water level recovers, and a mixture of sulphuric acid and iron in solution is created. This may discharge at surface with an ochreous stain as iron is precipitated out of the solution once it comes into contact with oxygen in the atmosphere (Fig. 8.5).

Metal mines may generate highly acidic discharges where the ore is a sulphide mineral. The predominant metal ion may not be iron but rather zinc, copper, or nickel. The most common ore of copper is chalcopyrite, which is a copper-iron-sulphide, which in the oxidation and hydration process produces a particularly toxic mine water.

Figure 8.5 Ochreous acid mine drainage contamination of a river below abandoned metal mines in Colorado. Image credit: Shutterstock/ Supitcha McAdam.

Acidic mine drainage can occur immediately the pumps are switched off, but it is usually a year or two before water levels recover to allow flow to surface. Generation of acid mine drainage may continue for decades or centuries after it is first detected, diminishing exponentially, and it is a serious long-term environmental threat.

Acid mine drainage is a growing problem in the world, especially in countries such as South Africa, with legacies of intensive deep mining. However, the technology exists to remediate the discharges and prevent them from harming surface water systems and from contaminating downstream groundwater systems.

Acid mine drainage treatment

A number of chemical treatments are used to neutralize the acidity and precipitate the iron and other metals from the mine discharge water. The most common is lime treatment, whereby a slurry of lime is dispersed into a tank containing the mine drainage and recycled sludge. This increases the pH to about 9, when most toxic metals become insoluble and precipitate to the bottom of the tank. Optionally, air may be introduced into the tank to oxidize iron and manganese and assist in their precipitation. The resulting slurry is directed to a sludge-settling vessel, so that treated water can overflow for release.

The most common methodology does not use chemicals, but rather creates a natural environment to treat the discharge water; this is the constructed wetlands system. The mine water is first trickled over limestone beds with full access to oxygen from the atmosphere. This raises the pH towards neutral, and some of the metals precipitate out of the discharge at this point. Final clean-up is done in a reed-bed wetland where most of the metal precipitation occurs due to oxidation at near-neutral pH, with complexation with organic matter, and precipitation as carbonates or sulphides. The latter result from sediment-borne anaerobic bacteria capable of reverting sulphate ions into sulphide ions. These sulphide ions can then bind with heavy metal ions, precipitating heavy metals out of solution and effectively reversing the mine-water generation process.

The attractiveness of a constructed wetland lies in its low operational cost. It is limited to some degree by the metal loads that can be dealt with (either from high flows or metal concentrations). The effluent from constructed wetlands is well buffered at between pH 6.5 and 7.0 and can be discharged into surface waters, where it is diluted. The precipitates are left in situ in the reed-bed and slowly buried and isolated from the water system as deposition continues.

This technique has been widely used in South Wales and in the East Midlands Coalfield and elsewhere in England. Pennsylvania and Appalachia also have numerous wetland systems. A typical example of passive treatment is the River Pelenna in South Wales, where a 7 km stretch had been polluted for a significant length of time by acid mine drainage from abandoned mine workings. This resulted in the river water becoming ochreous and unable to support any forms of normal aqueous habitat. A series of passive treatment wetlands were created to treat the discharges to the river, and these resulted in between 82 and 95% of the iron being removed and an increase in pH of between 1.0 and 1.5. A marked increase in the fauna within the river was noticeable thereafter.

Salinization

Another common form of groundwater contamination of the surface is salinization (Fig. 8.6). If groundwater is used for irrigation, and evaporation occurs from the waterlogged soil, salts will precipitate at surface. If the groundwater pumped to surface and sprayed on the ground has a significant sodium concentration, then the soil can become heavily alkaline, reaching a pH of between 9 and 10. This alkaline build-up causes the soil structure to break down and the soil becomes clogged and impermeable. The process is irreversible.

In areas of irrigation by groundwater, the increase in ionic concentration of the soil water can be prevented if the soil is well drained. Ditches 2 m deep and perhaps 100 m apart will assist in dispersing water that is held in the soil zone back to the aquifer. In areas with a shallow water table the ditches will also reduce the water level below the soil zone allowing aeration of the soil.

Under normal humid maritime conditions and in more arid climates any contaminated groundwater seeping to the surface will contaminate surface waters to a degree that is moderated only by dilution with the surface waters. Thus,

Figure 8.6 Saline soil caused by irrigation with mineralized groundwater, Al Ain, UAE. Image credit: Shutterstock/ Poonsab Harnphayom.

a spring discharge with a capture zone with high pesticide application to a shallow water table will discharge the more persistent pesticides and their daughter products into the surface water system. These organic compounds may be toxic to aquatic life at very low concentrations; any local stream water is likely to be already polluted, and the groundwater input only adds to the problem. For the most part, groundwater discharging to or applied on the ground surface is of low ionic concentration and only weakly polluted. This will not cause problems to the health of the soil or to aquatic systems.

9 Coastal aquifers and small islands – a challenge

Groundwater meets seawater

Assessment of an aquifer that has at least one side as the near-**constant head boundary** of the sea poses a number of difficulties. Firstly, there is an interesting dynamic between the fresh groundwater and the saline marine water where they interact. Seawater is more dense than fresh water and the fresh water will float over the seawater bounded by an **immiscible surface** within an aquifer to form the so-called 'seawater wedge'. However, this is not a stable front as the fresh groundwater is upwelling against it to discharge as best it can beneath the beach, whereas the denser seawater is tending to flow downwards against the front. Clearly an element of mixing will inevitably take place.

The Ghyben–Herzberg theory

The location of the front is given in broad terms by the **Ghyben–Herzberg theory**. The concept that fresh water is less dense than seawater and can float above it was recorded in the writings of Pliny the Elder (AD 23–79). Joseph Du Commun, teaching at the West Point Military Academy from 1818 to 1831, next identified the phenomenon, and the hydrostatic relationship between immiscible freshwater and saltwater bodies was independently investigated by Badon-Ghyben and Herzberg in the 1890s and early 1900s. The overall conclusion was that the ratio of the elevation of the water table (h_1) to the depth of the saline interface beneath sea level (h_2) is about 1:40 based on the relative densities of the two fluids (Fig. 9.1).

Practice has shown, for the most part, that the Ghyben–Herzberg theory has been found to apply only in part because the caveats of the theory require a uniform porous medium, an infinite homogeneous aquifer, and many of the other standard Theisian controls. Given the widespread occurrence of karst conduit systems in coastal aquifers that have developed at the present sea level as well as at lower relict sea levels, the application of the Ghyben–Herzberg principle is fraught with

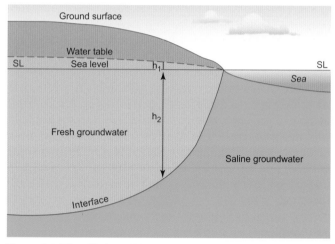

Figure 9.1 The Ghyben–Herzberg principal, h_1 to h_2 being approximately 1 to 40.

difficulty in many island situations. Conduit flow, for example, destroys the pressure balance, allowing ingress of seawater into aquifers to far greater distances than the simple saline wedge would allow. Obviously, the thickness of the aquifer is important; a very shallow aquifer will not allow ingress of seawater if there is sufficient head on the fresh water to the landward side to prevent this. A thick aquifer, however, under the same head will allow a steep-faced wedge of saline water to develop beneath the coast. In practice, the saline water to fresh water interface is a dynamic diffusion zone, and the ratio of the elevation of the water table to the depth of the middle of the diffusion zone beneath sea level is normally much less than 1:40, in some places even less than 1:20.

Island aquifers

Island aquifers are bounded on all sides by the sea, and are liable to saline intrusion from all directions. The importance of the dynamic mixing or diffusion zone, the so-called immiscible surface, was recognized in the islands of the Caribbean at an early stage. Electrical resistivity soundings on the island of Exuma were used to reveal the salinity curves that are characteristic of a shallow **freshwater lens**. The thickness of the mixing zone in North Andros and Grand Bahama, however, ranges up to 20 m, and this reflects the dynamics of the sea tides. Mixing can be enhanced by periodic imbalance of sea levels from one side of an island to another. Differences up to 0.3 m in sea level have been recorded across narrow limestone atoll islands in the Pacific as a result of prevailing winds. Under these conditions, the induced aquifer throughflow may cause significant perturbation at the saline water diffusion front. The saline interface here is a dispersion or diffusion zone that may be 20 m thick.

Dynamic dispersion models

The simple static Ghyben–Herzberg theory has now been superseded by dynamic dispersion models. These account for tidal lag and efficiency (time difference between and amplitude ratio of the water table fluctuations to sea tides), which vary with distance from the shore in low-lying limestone aquifers. In islands, vertical propagation from tides is greatest towards the centre of an island and horizontal propagation greatest near the perimeter (Fig. 9.2), although tidally influenced rises and falls in groundwater level are greatest near the coast. The available dispersion modelling codes are

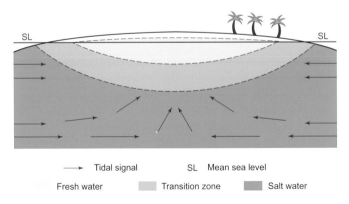

| → | Tidal signal | SL | Mean sea level |

Fresh water   Transition zone   Salt water

Figure 9.2 Dynamic dispersion model showing that vertical propagation from tides is greatest towards the centre of an island and horizontal propagation is greatest near the perimeter.

complex, but they replicate observed dispersion fronts and depths below sea level with considerable certainty. SUTRA, a code available from the United States Geological Survey, is the most commonly applied modelling software for use in small islands both in a non-tidal and a tidal mode. The model applies a 2D 'slice' solution across an island using generic dispersivity data.

Various rules based on the Ghyben–Herzberg principle are still applicable where the diffusion front can be approximated to a sharp line. Of these, the most common application is **Henry's Rule** whereby the highest elevation of the water table (H) in an island is proportional to the diameter, and in a long, thin island or sand bar, to the width.

Coastal aquifers

There is one other significant issue with regard to coastal aquifers. Whereas a mainland catchment water balance can largely be observed and measured, the water balance of an island is partly unseen and, therefore, almost impossible to measure directly. In simple terms, effective rainfall equals runoff plus the change to the groundwater store. River hydrograph separation allows these two components to be determined as runoff and groundwater baseflow. In the coastal context, surface water catchments may offer only ephemeral flow reflecting a recent rainfall event, and groundwater baseflow may largely occur not to the surface drainage system but directly beneath the beach to the sea or to coastal strips of mangrove (Fig. 9.3).

Simple estimates of the discharge per unit length of coast can be made using Darcian strip models, given some knowledge of the prevailing hydraulic parameters in the vicinity of the coast. Care is needed not to double account some of the groundwater discharge if the baseflow component is included in the surface-water runoff (streamflow). An island-wide groundwater flow model is the only way to establish the feasible range of the total groundwater baseflow.

A number of innovative techniques have been developed to quantify submarine discharges of groundwater, including measurement of flux indicated by thermal and chemical gradients beneath the sea bed, and mass balance of nutrients arriving offshore. A seepage meter can be deployed offshore by digging a ring into the sea bed connected to a deflated balloon, which inflates as water enters the apparatus from the sea bed. However, it is not clear that a point source discharge is always

Figure 9.3 Coastal mangrove, United Arab Emirates.

Figure 9.4 Groundwater draining out onto the foreshore at Magilligan Spit, Northern Ireland.

representative of a multi-layered system. Thermal imaging of groundwater discharges into the sea is also possible where there is a suitable difference in temperature between the sea and the discharging groundwater. Mass-balance calculations using natural tracers such as salinity and radon can also be applied to quantify the observed discharge plumes in the sea.

Resource management on small islands

Careful resource management is needed on small inhabited islands where demand can stress finite fresh groundwater reserves. Managers need to be informed how the groundwater system functions so that they can optimize the use of the resource and safeguard it from abuse. Although hydrogeological investigation in small islands is broadly similar to basin studies on the mainland, small islands of limited scale, coupled more often than not with data scarcity (including effective rainfall, permeability and baseflow) inhibit conventional groundwater flow modelling. Coastal and offshore baseflow measurement causes the greatest uncertainty. Small island hydrogeological investigations are challenging, and aquifers can be classified as high-elevation hard fractured rock systems and low-relief karst limestone and sand islands.

Groundwater discharge to a foreshore can often be seen at low tide, and specific electrical conductivity shows it to be a brackish mix of seawater and groundwater (Fig. 9.4). Even effective rainfall cannot readily be measured on some 'high-rise' islands where windward and leeward, coastal and interior may differ greatly and accurate monitoring is not practicable.

An extreme example is the rainfall gradient of 1000 mm year^{-1} km^{-1} between the coast and the upland interior on Raratonga in the Cook Islands. On the positive side, however, an island does have a coastline – to all intents and purposes, a near-constant head boundary – the maritime tidal signal generally only penetrates a few tens of metres inland except in karst limestone terrains.

Climate-induced sea-level rise

Climate change and projected seawater level rise will impact many of the low-lying islands and is perceived as a potential threat to the well-being of communities on low-relief islands such as atolls (Fig. 9.5). For the most part, islands will benefit from a slight increase in groundwater storage, but the seawater diffusion zone is likely to move inland. A rise in the water table reduces the thickness of the unsaturated zone and may even create wetlands in low-elevation areas. There is a downside to both situations: reduction in the unsaturated zone increases the vulnerability of the aquifer to pollution from surface and subsurface sources, and wetlands act as groundwater sinks in areas of high **potential evaporation** comparative to overall long-term rainfall. Some existing wetlands on tropical low-lying islands, for example Anguilla in the West Indies, are hypersaline due to evaporation.

Key issues on small islands are:
- management of demand and supply;
- freshwater resources on small islands are, by definition, shallow (vulnerable to contamination from on-site

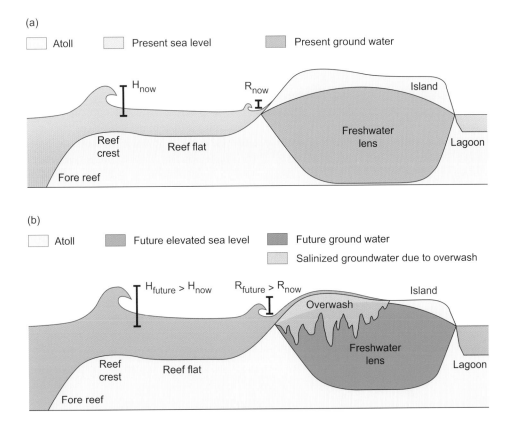

Figure 9.5 The effect of sea-level rise and wave-driven flooding on a low-lying atoll. (**a**) Current sea level. (**b**) Future sea level. Sea-level rise will allow for greater wave heights (H) and wave-driven runup (R), resulting in frequent overwash that will contaminate the atoll island's freshwater lens. Note: Heights are exaggerated. Image credit: United States Geological Survey.

sanitation, poor land-use practices and seawater intrusion);

- the Ghyben–Herzberg 'freshwater lens' theoretical model cannot be applied universally;
- runoff and groundwater discharge cannot be measured directly, as they would be with an inland catchment;
- impending sea-level rise will elevate base levels and change the freshwater storage capacity.

There are 50 000 small tropical islands in the Pacific, Indian and Atlantic oceans, of which about 8000 are inhabited, the majority to be found in the Pacific and West Indies. There are numerous small islands in the temperate zone, including, for example, several tens of thousands in the Baltic Sea alone, and 145 populated small islands around the coasts of the British Isles, although 20 of these have each fewer than 10 permanent residents. Many of these small islands enjoy high seasonal rainfall, yet are faced with severe water problems, especially on low-elevation coral atoll islands. It is noticeable that the high-elevation islands promote significant runoff, whereas many low-elevation islands have virtually no runoff. The tropical island paradox is human need competing with groundwater-dependent ecosystems; the water requirement to support subsistence crops such as coconuts, swamp taro, breadfruit and pandanus has a severe impact on the water budget. This paradox is repeated in temperate island communities where, for example, nitrate applied to early potato crops can cause long-term degradation of groundwater quality.

Small island typology

Studies of small islands inevitably draw on experience gained on larger islands such as Britain and New Zealand, which in turn draws on experience from continental studies, notably in North America, Australia and continental Europe. There are inherent dangers in this information chain because small islands cannot 'affect and be affected by atmospheric processes of the air masses moving over them', and are of a size where the time of response of even the larger surface-water catchments to a rainfall event is measured in hours rather than days, in an environment where every part of the island enjoys a maritime climate. The ratio of coast length to island area is a useful indicator of maritime influence. Other defining properties are island shape, maximum and average elevation, climate, runoff (perennial or ephemeral), groundwater transport and storage, vegetation and land use. Dynamic factors include sea-level rise, demand for water and engineered change.

A number of typologies have been published, some on a regional scale, whereas others have attempted a more universal classification. A typical generic typology is:

- geologically young 'high-rise' volcanic islands: Hawaii type;
- geologically older 'high-rise' volcanic islands: St. Helena type;
- near-continental bedrock islands: Channel Islands type;
- low-elevation coral limestone islands: Bermuda type;
- recent calcareous sedimentary islands: Turks and Caicos type;
- upland limestone islands: Malta type.

Groundwater vulnerability and islands

Population pressures

Small islands often suffer intense pressures. Although most support only small populations, a number are densely populated. Male in the Republic of Maldives, for example, has an area of just 1.3 km² but supports 60 000 people. Tourism can enhance water demand in parallel with the island's own needs; however, with the enhanced economy solutions such as seawater desalination become viable. Care needs to be taken to prevent pollution of the freshwater lens, and special care is needed in siting potentially polluting activities such as wastewater treatment systems, fuel stores, dry cleaners and laundries, the location of which might be condoned on a larger mainland aquifer with a deeper vadose zone and larger storage and diffusion properties. Perhaps the greatest threat of all is seawater intrusion, which can be avoided only where population density allows sensible groundwater management

based on sustainable supply; the most fragile case is a thin freshwater lens floating over seawater.

In some atoll communities the government has declared groundwater reserve areas in which settlement is prohibited and land use restricted. These are promoted in some Pacific islands by government payment to reserve landowners acting in their designated role as reserve custodians. At the core of many island groundwater management problems are traditional water ownership 'rights' and 'laws' inherent in land tenure. These rights conflict with the needs of urbanized island societies and take time to resolve through the introduction of appropriate legislation and subsequent administration.

Groundwater vulnerability

Groundwater in islands, especially low-lying limestone islands, is vulnerable both to natural processes and anthropogenic abuse. Critically, the water table is shallow and is vulnerable to chemical or biological surface pollutants. Even dry deposition of salt can arrive at the water table within two hours of the onset of rain after a long dry period.

Drought, storms and climate change can severely affect quantity and quality of the groundwater reserve. Periodic drought and lack of a reliable rainy season reduce the freshwater store as baseflow discharge and abstraction continue, albeit at a reduced rate. Salts continue to diffuse into the reserve and some of the fresher water may become brackish. Storm surges (and tsunamis) can overrun islands and pollute groundwater reserves with salt water and other contaminants. Uncertainties of future climate exacerbate the vulnerability of island groundwater resources. Sea-level rise is not a critical threat, as present-day storm surges can exceed predicted sea-level change estimates significantly in many regions. Indeed, small increases in base level may have a positive effect on many low limestone islands and induce a small increase in overall groundwater storage, although vulnerability to pollution may increase. Some small atolls, however, are under threat.

Extra care is needed to contain spillages from swimming pool chemicals, dry cleaning fluids, hydrocarbons, airport fire training foam and many other chemicals needed by island society. Many island nations, including some of the Marshalls and the Kiribati islands, have seawater reticulation for water flush sewerage systems. This is an attractive option, but care needs to be taken to avoid leaks and to ensure that the treated waste is not allowed onto the land. Treated waste water from freshwater flush systems (e.g. at tourist compounds) may often be used to water gardens, but both quantity and quality of the discharge need to be monitored to avoid mobilization of bacteria, nutrients and other chemicals.

Coastal dune fields

Dunes are an important feature of the British coastline. In England and Wales, for example, they often form shallow rain-fed aquifers draining to the periphery of the dune system – to the fore dunes and the beach as well as inland to low-lying land. The elevation of the water table is critical to the ecology of the dunes. Those with hollows or 'slacks' that are flooded in winter and dry out with a shallow water table beneath them in summer are preferred to slacks that are dry throughout the year (Fig. 9.6).

Coastal dunes typically consist of wind-blown sands, often including material from the exposed sea bed in the last glacial period. These sands have accumulated on top of impermeable estuarine or glacial clays, which in turn form the base of the dune system aquifer and provide a distinct hydraulic barrier to the underlying bedrock strata. The British climate provides an average annual excess of rainfall over evaporation that causes an accumulation of water within the dune systems, creating a so-called '**groundwater dome**', although the outflow conditions at the edges vary. The extent and height of the dome, and hence the groundwater head distribution

within the dune system, is controlled by the physical properties of the sand aquifer, specifically by the dune sand storage coefficient and the distribution of permeability or hydraulic conductivity. Values are often approximated from grain size analysis, as conventional test pumping methods cannot easily be applied in thin, high-conductivity aquifers where the transmissivity declines rapidly as the drawdown from pumping increases. Moisture content profiling can be applied to help improve understanding, by providing a recharge value against which formation constants can then be validated.

Geometry and spatial dimensions of the dune system are also important in controlling groundwater head distributions. For example, the elevation of the base of the aquifer relative to main drainage features controls the saturated thickness of the aquifer and hence its transmissivity. Other important factors controlling groundwater head distributions within dunes are the functionality of man-made drainage systems and the effective precipitation (i.e. the amount of water that infiltrates into the dunes following interception of rainfall on the plant canopy, recharge of any soil moisture deficit, evaporation from wet plant surfaces, and transpiration by vegetation).

The physical vulnerability of dune systems can be illustrated by a single storm event at Whiteford Burrows, a coastal dune spit wetland in South Wales. The present dune system is estimated to be about 600 to 800 years old; there are no pre-medieval remains within the sand although

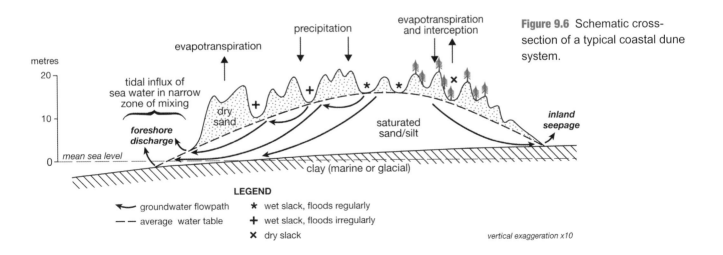

Figure 9.6 Schematic cross-section of a typical coastal dune system.

LEGEND

← groundwater flowpath   ★ wet slack, floods regularly

-- average water table   + wet slack, floods irregularly

✗ dry slack

vertical exaggeration x10

there is abundant archaeological evidence of early medieval activity nearby. The tidal range is large (c.7.5 m at spring tides) and offshore currents are strong. Whiteford Burrows spit is susceptible to change caused by periodic accretion and erosion of the beach; the height of the ridge of groundwater within the sand aquifer is essentially proportional to the width of the spit in accordance with Henry's Law.

A severe late winter storm event on 17 March 1995 caused extensive erosion of the foreshore, reducing the effective width of the dune system by 4%, so increasing the hydraulic gradient towards the beach (Fig. 9.7). The water table elevation fell immediately by up to 1 m. It was later shown that the erosive storm event caused a 6% decline in mean summer groundwater levels, a 1% decline in winter levels and a significant increase in drought magnitude, frequency and duration. The rapid water-level decline cannot readily be accounted for either by the lower rainfall recorded overall in 1995/6 or by the 'drought' of the mid-1990s. It was believed that this incident was a rare occurrence, but available historical evidence suggests that repeated periods of erosion and accretion have occurred over at least the last 200 years at this particular dune system.

Each dune system has its own unique drainage characteristics (e.g. direct flow to the sea, free drainage inland, agricultural drains, controlled water levels or adjacent river with varying stage or water level). Discharge to the beach beneath the fore dune creates a diffuse brackish zone rather than a saline wedge. This zone is flushed twice daily by sea tides.

Land use varies between the systems. Some are actively grazed, whereas others are left un-managed or have partial tree cover. The sensitivities of the water tables may differ; one site may be susceptible to changes in vegetation, another to coastal erosion and accretion or to variations in the rainfall regime. It is important to understand these sensitivities in order that each coastal dune system can be managed to best advantage.

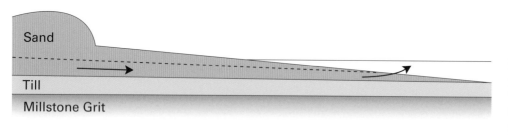

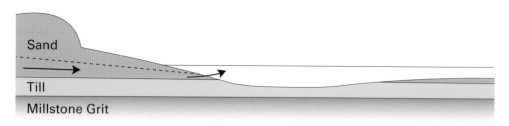

---------- Potentiometric surface (water table)

————— Schematic groundwater flow path

Figure 9.7 Storm damage at Whitford Burrows, South Wales, causing an increase in the hydraulic gradient out of the dune aquifer towards the beach.

10   Flood, drought and subsidence

Groundwater flooding

There are five key types of **groundwater flood** event:

1. natural rise in the water table caused by extreme high-intensity and/or long-duration rainfall, resulting in surface water ponding, intermittent stream flow or the anomalous activation of springs;
2. groundwater flow in alluvial deposits by-passing river channel flood defences;
3. rise in the groundwater level owing to cessation of groundwater abstraction;
4. underground structures creating barriers to groundwater flow, resulting in water tables rising to cause flooding;
5. **turloughs** (Ireland) or poljes (Balkans) in karst limestone terrain.

All of these mechanisms can result in significant surface water flooding. The first of these is called a groundwater flood event, where groundwater may emerge from either point or diffuse locations.

Groundwater flooding

A groundwater flood can occur over specific parts of the Chalk aquifer in England and elsewhere, but is not commonly recorded on other unconfined aquifer systems such as sandstone aquifers. Although the exact nature of the processes that cause groundwater flooding remains uncertain, the overall causes of these events are understood. Evidence from flood events at Chichester and Patcham in the English South Downs aquifer and from the Pang–Lambourn catchments in the Berkshire Downs indicates that there are two types of groundwater flood event. Type 1 is the true groundwater flood in which the water table elevation rises above the ground surface, and Type 2 occurs when intense groundwater discharge to a **bourne**, ephemeral spring or a highly permeable shallow horizon discharging to surface waters causes overbank flooding. A bourne spring starts to flow in a dry chalk valley as the water table rises up the length of the valley during intense wet weather (Fig. 10.1). It will cease to flow as the water table recedes.

Figure 10.1 Bourne type spring at the head of the stream flow in the Lambourne, in southern England, just below the point where the water table intersects the stream bed. Image credit: Jeff Davies.

Groundwater flooding depends on the wetting up of the soil over a sustained period, where there is no soil moisture deficit. If this is accompanied by prolonged rainfall recharge at a rate that exceeds the drainage potential from the aquifer at a local or catchment scale, then flooding is likely. The water table rises until it intersects the ground surface, causing groundwater effluent flow into an otherwise dry Chalk valley.

Flooding in alluvial deposits

A major flood event at Chichester in Sussex, southern England, occurred when the River Lavant overtopped its banks in the winter of 1993–1994. Groundwater levels rose in response to higher than normal autumn rainfall, and then in December and the first half of January the area received an additional 350 mm rainfall. The water table rose in response to these conditions and caused numerous episodic springs and seepages to appear along the valley of the River Lavant,

West   East

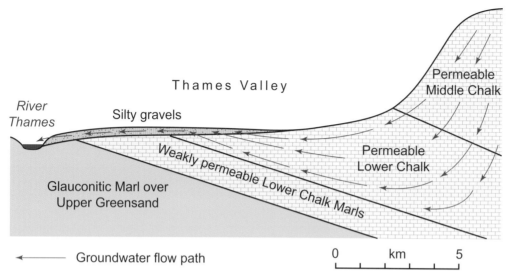

Figure 10.2 Cross-section of the Chiltern Hills chalk scarp showing discharge of groundwater to riverine gravel – vertical exaggeration x2.

Thames Valley

Permeable Middle Chalk

River Thames

Silty gravels

Permeable Lower Chalk

Weakly permeable Lower Chalk Marls

Glauconitic Marl over Upper Greensand

◄─────── Groundwater flow path

0   km   5

and significantly increased the river flow. At the same time the river response turned flashy, with any new rain event dispersing as surface runoff, the soil now saturated, and the Chalk aquifer below it effectively fully saturated as well.

Local flooding of alluvial lowlands can be caused when the water table intersects the ground surface, and this is controlled essentially by local river stage. In the Thames valley between Reading and Oxford, groundwater flowing down from the Chiltern Hills scarp slope within the Chalk aquifer discharges along a spring line within the Lower Chalk (Fig. 10.2). The spring line occurs at the diffuse junction between the permeable fractured chalk and the marls towards the base of the Lower Chalk. These marls act as a confining layer to the underlying Upper Greensand aquifer. The chalk groundwater is forced upwards to discharge as springs; some of these are ephemeral while others flow throughout the year. There are also numerous underground discharges into the First Terrace Thames Gravels, a silt matrix granular deposit of moderate permeability that rests on top of the Chalk in the Thames Valley near Wallingford. The gravels normally discharge the received water and direct rainfall recharge down hydraulic gradient directly into the River Thames, but when the river is overbank and the Chiltern Hills are thoroughly wetted up, spring discharge to surface is increased (Fig. 10.3). However, the groundwater in the gravel is then backed up against the

Figure 10.3 Flooding from the ephemeral chalk spring at Crowmarsh Gifford during 1941. Image credit: David Beasley.

high river stage. Flooding from the spring discharges will occur, exacerbating the overbank flooding of the river. In addition, localized surface flooding pools may occur wherever the gravels are low-lying.

Exceptional rainfall

Groundwater emergence, as the water table rises above ground level, has occurred at Patcham, in the northern suburbs of Brighton on the south coast of England. The Borough archives record flooding for the years 1877, 1913, 1915, 1916 and 1925, the period 1913–1916, which represents a potentially cyclic return period of intense wet weather, and the winter of 2000–2001. The severity and extent of the 2000–2001 event reflected both the meteorological conditions and high **antecedent** groundwater levels. In the period from September 2000 to April 2001 there were eight successive months of exceptional rainfall. As a result, some southern areas of the United Kingdom received double the long-term average (1961–1990) rainfall during September–December 2000 and there was no precedent for the intense recharge the aquifer received that winter.

The likely mechanism of the surface flooding at Patcham was simply the saturation of the Chalk matrix, the water table rising until the entire system was saturated between the normal water table elevation and the ground surface. The result was surface ponding with the flood waters characteristically blue-grey in colour, standing up to 1 m deep in some places. The blue-grey colour typical of Chalk water discharges contrasts with the turbid brown coloured waters that are characteristic of overland flow. The colour thus indicates that the water has risen out of the ground rather than as a surface water flow picking up suspended solids on the way.

The groundwater flood scenario related to bourne-type spring discharge recorded on the Pang and Lambourn sub-catchments to the Thames in Berkshire is recurrent. There is a history of incidents dating back to the 1930s. Two significant groundwater flooding events have occurred recently; the first during the winter of 2000–2001, at the same time as the Patcham flood, and a smaller event that occurred in the winter of 2002–2003. Both coincided with unexceptional and brief rainfall events, not normally associated with surface flooding, but following prolonged and intensive wet weather. Investigation of the Pang–Lambourn flood events indicates that when the aquifer receives about 100 mm of effective rainfall per month for three consecutive winter months, it attains a potential to flood parts of the catchments. Additional rains dictate the length and severity of the flooding. Hence in 2000–2001 the prolonged high effective rainfall continued after the initiation of flooding, and high water levels and associated flooding persisted for about eight months, whereas in 2002–2003, the water levels receded about two months after the monthly rainfall dropped below average in February. The two catchments are typical Chalk bournes whereby stream flow, in the otherwise dry valleys, starts at a spring source that reflects the elevation of the water table. Thus in winter the stream will generally flow from a higher elevation within a bourne catchment than in summer.

A turlough or polje is quite different. Point source flooding in flat-lying karst limestone may occur in sink holes in a slight low-lying depression, which as the groundwater level rises in the wet season become emergent and flood the surrounding land. This phenomenon is known in Ireland as a turlough or dry-lake. Turloughs are mostly found on the central lowlands of Ireland west of the River Shannon, in counties Galway, Clare, Mayo, and Roscommon. There is one also in South Wales. Most turloughs flood during the autumnal rains, usually some time in October, and then dry up between April and July. However, some turloughs can flood at any time of year in a matter of a few hours after heavy rainfall, and they may empty again a few days later. Some turloughs are affected by marine tides: in the summer, a turlough situated 5 kilometres inland from Galway Bay floods and empties twice every 24 hours. Most turloughs flood to a depth of about 2 m but some are deeper. In the Dinaric karst limestone the term polje is used instead of turlough (Fig. 10.4).

All of the Irish turloughs are found in Karst limestone areas. To the east of the Shannon, the limestone is often covered by a thick cover of glacial drift, but in many areas to the west of

Figure 10.4 Planinsko polje is a typical karst polje in the Dinaric karst, Slovenia. The lower-lying land floods with groundwater at times of intensive rain. Image credit: Shutterstock/Stepo Dinaricus.

83

the Shannon, where the limestone is pure calcareous rock and the drift cover is thin, there is no obvious surface water drainage network. In these areas, rainfall passes underground, flowing through cracks and cave systems in the rock to rise at distant springs. In winter, when groundwater levels rise and the underground flows increase, and when the conduits to the springs are not capable of dealing with the amount of water entering them, groundwater may appear temporarily at the surface.

Engineered drainage

Not only is groundwater flooding at the surface an issue, but groundwater can also be a common problem in subsurface workings such as quarries, opencast sites and even building sites. Dewatering behind a rock face, however, provides an element of stabilization by reducing the **hydrostatic pressure** within any cracks and fractures. Whenever the construction engineer excavates below ground level there is some danger of penetrating below the water table and encountering ground-water problems. Hazards may include inflow of water, heave of excavation base and collapse of soil material. This is par-ticularly the case in countries that have a generally shallow water table and extensive areas of recently deposited allu-vium or glacial clay. Most of the industrialized countries of the world are affected in this way; groundwater problems are, therefore, a worldwide hazard for excavations, and engineered dewatering is in universal use.

Groundwater problems sometimes come as a great surprise to the construction engineer, who will immediately classify them as 'unexpected ground conditions' with consequent claims being made against the employer. Good ground investigation is thus an essential precursor to any sub-surface engineering work.

There are a number of engineered options to inhibit ingress of groundwater to a sub-water table site:
- Removal of groundwater and relief of pressure:
  - Sump pumping
  - Groundwater interception by boreholes or **well points**
  - drilled wells, relief wells and galleries
- Exclusion of groundwater:
  - Sheet piling or diaphragm walls
  - Slurry trench cut-off or thin grout membrane
  - Contiguous bored pile walls

- Cement, clay, resin or silicate-based grouts
- Ground freezing
- Compressed air

The subsequent abandonment of engineered barriers will reinstate natural groundwater conditions except where dia-phragm walls, grouting or other cut-offs remain. Foundations that were laid in dry conditions thus become saturated and **borrow pits** flooded.

Groundwater removal

Removal of groundwater by sump pumping is an effective way of dewatering areas such as quarries and mine workings. The water removed from the workings, however, may be contaminated by the processes surrounding the sump. In order to avoid this, a ring of boreholes around the quarry or mine can be used to intercept groundwater flowing towards the excavation. The boreholes need to be greater than the depth of the excavation in order to be effective, and their combined 'cones of depression' need to encompass the whole periphery of the workings. This allows removal of clean groundwater, which may then be used as process water.

Well points work on the same principle but are relatively shallow and closer together and are often deployed in soft ground surrounding a construction excavation in an area of shallow water table (Fig. 10.5). They are commonly used for building site work close to the sea in flat-lying coastal areas, in order to exclude seawater, for instance on the coast of the United Arab Emirates. They are also applied to trench work – for example, laying buried pipelines. They are simple to install and can be let into the ground by passing a high-pressure jet through them as they are knocked in. This also grades the granular material surrounding each well point.

Groundwater exclusion

Exclusion of groundwater is exactly that: the building of a sub-surface, vertical impermeable barrier to prevent the passage of groundwater through the barrier. These measures may be temporary, such as ground freezing, or may be per-manent when applied to embankments or dams and other hydraulically sensitive structures.

Groundwater rebound

Development of the steam-driven drilling machine in the late-nineteenth century, coupled with intense demand for water

Figure 10.5 Part of a groundwater interception well point system protecting an excavation. Image credit: Shutterstock/Bildagentur Zoonar GmbH.

in industrial areas and cities, caused rapid decline of water levels in many aquifers. Water levels remained depressed until the decline in industry from the 1970s onwards and in many areas, notably in England at London, Birmingham, Liverpool and Manchester. The water table then slowly began to recover for the first time in over one hundred years. Rebound was enhanced by additional recharge from cracked and broken Victorian sewers and from broken water mains.

In the long period of depressed water levels, many subsurface structures had been built, for example, the London Underground train system, in the belief that the water table would remain at its dynamic depressed level indefinitely. In Central London until about 1820, groundwater levels were about 5 m above Ordnance Datum and the Tertiary sands beneath the London Clay were naturally fully saturated. Groundwater abstraction thereafter induced a steady decline in water levels so that the sand was dewatered sometime between 1850 and 1875. More recent years have seen a reversal of this trend, and water levels have recovered steadily, so that resaturation of the sand and clays is now ongoing. This causes upheaval of the clay as pore pressures increase in the sand and overlying clay. The problem is not confined to the United

Kingdom, and both the New York Transit Authority and the Paris Metro have experienced flooding. In all these instances expensive engineered remedial action has been required.

There are also problems of dewatering where tunnelling has taken place below the base of the clay and into the Tertiary sands. These sands are, in places, pyrite-rich, and the resaturation of the oxidized pyrite can allow iron and sulphate to be taken into solution to create an acid mine drainage type water.

A different problem has arisen in some Middle Eastern cities, which has caused saturation of the shallow aquifer system that overlays an extensive **aquitard**. In Riyadh, for example, high water demand was met by long-distance imports of desalinated water. The average water consumption in the city of well over one million people is an extravagant 600 litres per person per day. However, once imported water was on line, pumping ceased from the deep limestone aquifer beneath the city that had become contaminated. Within time significant water leakage occurred, derived from the new desalinated water supply: from water mains and storage tanks, sewers and septic tanks as well as from intensive irrigation of open parklands. The outcome was extensive water-logging as the water table in the shallow aquifer reacted to this new form of recharge, and the water table soon intercepted the ground surface. The flooding of basements and damage to underground facilities has been extensive but drainage of the aquifer has been ineffective. Remediation finally came about when renewed and intensive pumping in the deep limestone aquifer was found to slowly drain down levels in the shallow aquifer above the aquitard. Similar problems have also arisen in both Kuwait and Doha.

Groundwater drought cycles and drought proofing
Historically, around 20% of the land surface of the planet experiences drought at any one time. This has now risen to 28%, and is set to rise to 35% in the early 2020s due to changing climate. In recent years, areas affected by the most severe droughts have risen from 1% to 3% of the overall landmass. In many parts of Africa, for example, the occurrence of drought is now considered the norm and not the unexpected. There is considerable uncertainty surrounding the future of Africa's climate due to climate variability: decreased rainfall in

northern and southern Africa, increased rainfall over the Ethiopian Highlands and (most notably) the increased frequency of both floods and drought.

Drought cycles
Rainfall within the semi-arid region of Southern Africa is seasonal and episodic with distinct wet and dry cycles. An 11-year long cycle over Malawi, Tanzania, and Zimbabwe increases in duration to the south, where a less distinct 18-year cycle is prevalent in South Africa (Fig. 10.6). Groundwater recharge mainly takes place during the wet cycle, a period of five or more years when rainfall sustains crop growth, animal husbandry, and family wellbeing. During the dry climate cycle, communities are solely reliant on groundwater storage.

In many parts of the semi-arid region, supply in the past was usually sustained to a greater or lesser degree, with some hardship experienced towards the end of the longer dry climate cycles. Typical modern long-term average recharge estimates in the region are in the range 0.2 to 13 mm/year,

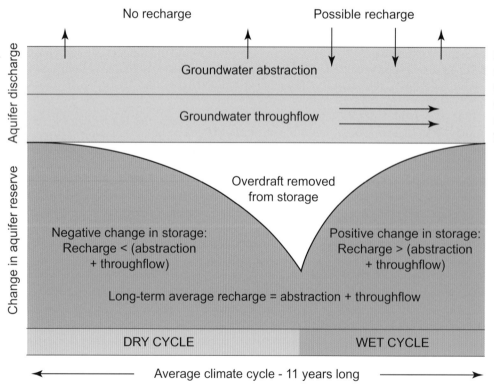

Figure 10.6 The groundwater resource and the drought cycle, showing progressive depletion during the dry period and resource recovery in the wet years.

while in Malawi they are between only 0.1 and 4.7 mm/year. Demographics and greatly increased demand for water have been partially satisfied by large drilling programmes. For example, the Government in Malawi reports a six-fold increase in the number of boreholes in the country since 1980, but their impact is moderated as this increase includes short-term drought relief and refugee resettlement drilling programmes.

Reliable statistical analysis of source and resource failure are difficult to come by, as little long-term monitoring of groundwater abstraction rates and groundwater levels is undertaken in much of arid and semi-arid Africa. Available reports strongly suggest that recently increasing numbers of resource failures reflect over-abstraction of the groundwater resource, rather than mechanical or structural failure of pump or borehole, regardless of the age and previous sustainability of the supply. A significant percentage of new rural water-points completed in Malawi in the last decade, each comprising a borehole and a hand pump, have failed within three years of being commissioned. Many of these failures reflect periodic drying-up of the resource.

But drought is not just an African problem, and periodically occurs in many parts of the world where problems are now exacerbated by the vagaries of changing climate. A similar story of aquifer depletion is coming from parts of India, where over-pumping is also drying up wells in the crystalline basement rock aquifer.

Groundwater will persist longer than surface water at the onset of prolonged dry conditions, and will provide a reliable supply long into a drought event (Fig. 10.7). However, it takes much longer to recover at the end of a drought, when rivers and lakes have been replenished by rainfall.

Planning for drought

Identification of areas susceptible to drought is a valuable means of deploying resources to 'drought proof' an area with additional deeper boreholes and improved surface storage facilities. **Groundwater drought vulnerability maps** do just this by combining three basic layers of information: groundwater resource potential, prevailing rainfall and rainfall distribution patterns, and population distribution. The information is weighted according to simple algorithms. An area that is poorly populated and has a low groundwater resource potential is not as susceptible to drought as a moderately endowed groundwater regime that has a large water demand from a more densely populated area. These three generic data levels can readily be combined in GIS format to produce simple groundwater drought vulnerability maps at scales that can help target drought proofing strategies. In addition to the three basic input layers, drought vulnerability maps can be refined to encompass a range of additional data. These include, for example, community dependence on groundwater with specific regard to livelihoods, land use and environmental resilience to drought, existing coping strategies, drought preparedness and Government's capacity to implement timely drought intervention.

Climate variability
Change in resource capacity

The potential impacts of climate change on groundwater systems are poorly understood due to the complex relation that exists between groundwater and climate variables. For example, recharge depends on the timing and quantity of precipitation, and whether it falls as rain or snow. Evapotranspiration is dependent on quantity and timing of precipitation as well as air temperature, wind speed and direction, humidity and solar radiation. But other influences may evolve at the same time: anthropogenic activities, including groundwater abstraction resulting in decline in storage, and the capture of natural groundwater discharge, which often accompany some elements of climate change. Thus, distinguishing between anthropogenic and climate-induced changes in groundwater systems is difficult. For example, the magnitude and phase relation of the El Niño-Southern Oscillation, Pacific Decadal

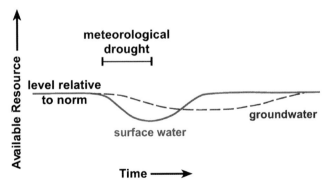

Figure 10.7 The reaction of groundwater to drought is slower than surface water, but also slower to recover post-drought.

87

Oscillation, and Atlantic Multi-decadal Oscillation **climate cycles** may result in average or extreme climate conditions that may affect precipitation, drought, infiltration, recharge, discharge, storage and quality of groundwater resources.

Projected changes in precipitation due to climate change display spatial variability indicating positive and negative impacts on both surface and groundwater processes. Projected warming trends will significantly affect global and regional evapotranspiration patterns, with direct implications for the sustainability of surface water and groundwater resources. Higher air temperatures will clearly increase evapotranspiration and reduce runoff and soil moisture, thus adversely affecting recharge potential.

Within the semi-arid to savannah areas of southern Africa, climate change may produce drier conditions due to reduced average annual rainfall. As a consequence, drought periods may become increasingly prolonged. The wet parts of a 20-year drought cycle may result in only very little rainfall recharge to aquifers because evaporation is greater than rainfall for much, if not all, of the year. In such circumstances, aquifer recharge may only really take place during infrequent tropical cyclonic storm events. In areas already experiencing such climate patterns, e.g. southern Zimbabwe, northern South Africa, eastern Botswana, Mozambique and southern Malawi, there will be an increased reliance on groundwater for domestic water supply, especially during drought periods, and therefore, application of meagre resources to the cultivation of crops by irrigation should be discouraged.

Small changes in precipitation and/or evapotranspiration may lead to large changes in soil moisture and recharge in many semi-arid and arid regions, although in some other regions groundwater recharge is likely to increase with a rise in precipitation. The proportion of land surface that will experience extreme drought is predicted to increase, especially in the sub-tropics and mid-latitudes. Different aquifers and different locations within a given aquifer could experience varied climate-induced changes, depending on their hydraulic properties. Many studies indicate substantial alteration in the hydrologic cycle in snowmelt-dominated regions through seasonal shifts in stream flow.

Change in quality of resource
Climate variability may also impact groundwater quality. Changes in recharge rates, mechanisms and locations will affect contaminant transport. Climate change may alter the timing or the conditions governing chemical interactions; this could degrade the quality of the groundwater. Also, the spatiotemporal variability in precipitation patterns may result in intense and prolonged infiltration events. Pore-water salt reservoirs in the vadose zone, mainly chloride and nitrate, could be flushed into many aquifers leading to increased level of salts in groundwater.

A deeper understanding of the effects of climate change on groundwater resources over the long term is integral for better planning and management of groundwater resources. However, hard information about climate-related effects on groundwater resources is currently inadequate, especially with respect to groundwater recharge, discharge, quantity and quality. Knowledge of climate-induced impacts on groundwater systems is limited by uncertainties inherent in the assessment processes.

Land subsidence
Sedimentary formations are formed as soft sand, silt and mud. As the sedimentary pile increases in depth, so the weight of overburden compresses the deeper beds until the skeletal structure plus the hydrostatic pressure within it balances the overburden pressure. Groundwater abstraction from such sediments reduces the pore-water pressure and increases the effective stress from the overlying strata. Ultimately the system gives way, at a point known as the **pre-consolidation stress point** and sediment compaction occurs, accompanied by land subsidence.

Coarse-grained sandy aquifers form a rigid aquifer framework that generally resists compaction, whereas fine-grained clay-bound strata are more plastic and more prone to compaction. Where the two are sandwiched together to form a sequence of aquifers and aquitards, groundwater abstraction from the coarse layers can cause leakage from the finer-grained layers, resulting in their compaction. This can cause serious land subsidence at the surface, as experienced in Beijing, Bangkok and Jakarta. In places this has allowed high tides to overwhelm the land, causing salinization and extensive flood damage to urban areas.

Sink holes
A **sink hole** can occur rapidly without warning through the collapse of overburden above a cavity. The cavity is usually the

Figure 10.8 A sink hole in dolomite in Gauteng, South Africa. Image credit: Jude Cobbing.

result of groundwater dissolution of soluble rock such as limestone or dolomite (Fig. 10.8). Collapse can be life threatening if it occurs in an urban area. On the Chalk of England weathered sink holes, rounded depressions in the ground surface, are called **dolines**. They are typically up to 10 m in diameter and 5 m deep.

11 Some topical issues

Overview

Climate change is undoubtedly the major issue impacting on groundwater at the present time (Chapter 10). This impact may result in both a decline of recharge and deterioration of groundwater quality. Rising sea levels will raise the base level of rivers and streams as well as coastal and island aquifers (Chapter 9).

Apart from climate change there is the thorny problem of how the hydrogeologist informs the managers and politicians and educates the public at large. It is now urgent that we spread the word globally that groundwater is crucial to the survival of the planet, that it must be protected, safeguarded and used in a sustainable manner, and that use should be allocated equitably in proportion to the perceived social and economic value of the type of demand, whether domestic, agricultural or industrial. There has been much criticism of scientists talking to scientists in learned papers, and of hydrogeologists preparing maps that only they can understand. This failing on the part of the scientist needs to be addressed, and communication needs to become a part of the hydrogeologist's daily routine. This book is one small part of that process.

Geological disposal of waste

There is also public concern about the major and minor ways in which groundwater flow allows migration of radionuclides or gases. Nobody wants a nuclear **waste repository** 'in their back yard'. However, the British Government geological screening process of local authority areas in England and Wales has brought positive results. Local area authorities that recognize the economic benefits of a repository being constructed in their area have come forward for screening. The public perception is that the geological disposal of nuclear waste could be a danger to public health. The reality is that provided the right investigatory work is carried out to evaluate groundwater flow regimes, their provenance and emergence at surface with very long transit times, then a safe repository is feasible.

Oil-shale fracking

There is a similar public perception concerning deep oil-shale fracking for gas, a process that has been ongoing in the United States since the mid-twentieth century. There have been accidents, but for the most part the main contamination risk to nearby groundwater bodies is the arrival of small quantities of methane coming up from the fracked shale beds. Again, provided that the right kind of investigation is thoroughly conducted before work begins, prevention of groundwater contamination is possible.

Small-scale issues

Other smaller-scale issues are topical or becoming topical. One of these is groundwater **heat source pumps**. These use the heat or thermal energy stored in groundwater to warm buildings in winter and cool them in summer. The concept is that the energy exchange is sustainable and renewable because of the size of the groundwater body compared with its contact with a single heat pump loop. On a city scale, a problem can arise when neighbouring heat pumps start to interfere with each other. This was the case in London when London Underground Limited planned to adopt heat pumps to cool its railway stations. For the most part this was impractical, as neighbours had already swamped the heat exchange potential in the Greensand aquifer.

There are many other local issues. These include contamination from cemeteries, not only organic contamination but metals and other inorganic pollutants. A common pollution source is salt for application on roads and pavements during periods of frost. Unless the stores are properly engineered, saline runoff can quickly contaminate groundwater, and salt-rich runoff occurs following application to surfaces.

Distribution of resources

One of the more difficult political issues that we hear about is the apportionment of groundwater resources between neighbouring states – no easy task, given the diffuse nature of groundwater flow. Nevertheless, with the appropriate investigation and the gathering of time series data on water levels and quality, this can be done, provided there is a political will to do so. One example fraught by such difficulty is the equitable apportionment of the limestone Mountain Aquifer between Israel and the West Bank. The political element becomes extreme when groundwater resources dry up due to mismanagement or over-use. The outcome can often be violence, reverting in places to tribal factions, one against the other.

Global pressure on resources

Of the many other public concerns around potential hazards to groundwater, some are global; some only impact the arid and semi-arid climates, while others are entirely aquifer-specific. But despite all these concerns, groundwater continues to serve its dependent communities, including the many major cities around the world that have no other source of supply. These include Mexico City with a population approaching 27 million, Calcutta with nearly 17 million and cities such as Dhaka, Manilla and Cairo each with 12 million people dependent on groundwater. Over half of the so-called megacities, with populations greater than 10 million, are dependent to some degree on groundwater, including London. Urban groundwater dependence is greatest in China and elsewhere in Asia, in Central and South America, in Russia and in Africa. The key issue in all these instances is the protection of groundwater sources and recharge areas from the consequences of urbanization in terms of pollution and increased runoff.

All these issues are compounded by the ordinary threats of groundwater pollution (see Chapter 8) and over-abstraction that occur throughout the world. Groundwater remains a viable and essential source that underpins the global economy and the wellbeing of people not only in rural areas but in many urbanized areas and many of the larger cities.

Getting the message across

It is extremely difficult to get the message across to politicians and the wider public that groundwater is a valuable resource and needs to be nurtured as well as exploited sensibly. There is a universal reluctance to visualize groundwater – 'it cannot be seen or heard so, as far as I am concerned, it is not on the radar'. An engineer will look to surface water as a supply before he has to even consider a groundwater option. 'When did the Queen last open a borehole?', he will say, thinking back to the great celebrations accompanying the commissioning of a new dam.

During the 1990s the public water supply system in Northern Ireland retrenched almost entirely from its network of high-yielding borehole sources in favour of surface water sources. This was done despite the vulnerability to drought of upland storage sites and the near absence of any security of these resources from wanton damage. The reason given was that the boreholes were expensive to maintain, requiring electricity to pump the water up out of the ground, whereas the reservoirs only needed gravity to drain supply to the urban consumers. It did not take long before this policy backfired, and in 2018 Northern Ireland was the first area in the British Isles to declare a hosepipe ban because of the prolonged (hot) dry summer. Engagement with the managers and politicians to explain that abandonment of their groundwater sources was unwise was extremely difficult, and it was apparent that there was a gulf between science and the best interest of the public.

Addressing this gulf has become an increasing concern worldwide. In Africa, for example, the sight of domestic animals drinking from standing water around a village hand pump is commonplace, as also is the picture of women gathering drinking water from unprotected groundwater sources (Fig. 11.1). But common sense should dictate that animals and their faeces should be kept away from well heads and boreholes lest they contaminate the source. For the same

Figure 11.1 Women collecting water from an unprotected groundwater source in Ethiopia.

reason, common sense also dictates that pit latrines and soakaways should be distant from water points. There are a number of simple, freely available guidelines to help address these issues, but widespread ignorance prevails regarding the need to protect both individual groundwater sources and the overall resource.

Nowadays simple means have been adopted to get the message across, one being the use of cartoons, and another use of travelling puppet shows performed in rural communities. Other tools that have been developed include guidelines amply illustrated with simple graphics, as well as maps of various groundwater features. None are entirely satisfactory, and the hydrogeologist needs to develop improved communication with the wider public and with its leaders. This is a longstanding failure that has no easy answer. People cannot see groundwater, so why should they care about it? Getting them to even think about groundwater is a major challenge, but an essential and important one.

Radioactive waste disposal

To date, all the proposed waste repository sites are located below the water table, with the exception of the Yucca Mountain site in Nevada, where the proposed repository is in the unsaturated zone. The apparent dryness of the unsaturated zone is complicated by a high infiltration rate, the large volume of water that is held within the porosity, and rapid flow along fractures. For disposal below the water table, the host rock must have a low permeability to ensure that radionuclides released from containers over time do not reach the biosphere. Three types of host rock are widely recognized as having the potential to provide a sufficiently low permeability to meet this safety requirement:

1. strong rocks with very low porosity and very few fractures, including igneous, metamorphic, and certain types of sedimentary rocks;
2. relatively weak mudrocks or clays, which will not sustain open fractures for extended periods of time;
3. rock salt, which is also self-sealing.

Geological barriers exploit the properties of the rock and hydrologic system around the repository and are intended to extend the travel time for radionuclides to return to the biosphere. The movement of fluid through the rock is controlled by its hydraulic conductivity, which depends on matrix permeability and the presence or absence of fracture

systems. Irrespective of the movement of groundwater, the mobility of the radionuclides in solution may be further retarded by dilution, sorption onto mineral surfaces, precipitation of secondary phases that contain radionuclides, and matrix diffusion.

Constructing a repository for hazardous waste

To be suitable for the construction of a repository, the rock must possess certain physical properties. In some cases, such as granite, high mechanical strength favours construction, but the presence of fracture systems may increase the hydraulic conductivity. In the case of salt, a medium level of mechanical strength may be advantageous because plastic deformation and flow will seal the rock around the waste packages, each of which is contained in an engineered barrier.

From the perspective of geological disposal, the four critical issues for developing a repository are:

1. the radionuclide inventory;
2. how this inventory changes over time;
3. some sense of the geochemical mobility and radiotoxicity of the radionuclides;
4. the thermal output of the waste.

Storage over time

The challenge for geoscientists is to understand how the evolving conditions of a repository affect the geochemistry and mobility of the radionuclide inventory, which is also changing over time. The decay of one radionuclide often leads to the accumulation of another: for example, ^{239}Pu (plutonium), which has a half-life of 24 100 years, decays to ^{235}U (uranium), which has a half-life of 700 million years. Many proposed deep-mined geological repositories are intended for the disposal of heat-producing spent fuel or high-level waste. But within each of these waste types there are important variations. For spent nuclear fuel, the type of fuel, burn-up, and age since removal from the reactor all have an impact on the radionuclide inventory and the thermal output. For high-level radioactive waste, the type of chemical processing and the age of the waste are important.

Reprocessing generally lowers the content of long-lived actinides, but leaves high concentrations of shorter-lived fission products, such as ^{137}Cs (caesium) and ^{90}Sr (strontium). These have half-lives of approximately 30 years and a large thermal output, which in turn determines how long the waste must be stored at the surface before disposal. These complexities mean that countries with a long history of developing nuclear power plants and weapons programmes have a wide variety of waste that requires different treatments before contemplating disposal.

Oil shale and groundwater

Gas embedded in shale rock formations deep below the Earth's surface has long been considered inaccessible. This was due to high drilling costs and because shales lack sufficient natural permeability for the recovery of gas at rates suitable for large-scale production. Deep borings must be used and fractures must be engineered to enable commercial viability. New horizontal drilling methods, combined with innovative techniques to fracture the rock, have made shale gas production economically viable. New technology for gas production from shale formations was evolved in the Barnett Shale in Texas, and its economic success has led to the rapid exploration of shale formations in many countries, and has greatly increased the estimates of global natural gas reserves in the world. Not least, its production is helping to maintain stable energy prices.

Oil-shale fracking

Hydraulic fracturing, or fracking, in combination with horizontal drilling, is an essential part of the shale gas production process and has been in commercial use in the USA since 1948. Extraction involves drilling of deep horizontal wells and enhancing the natural permeability of the shale by **hydraulic fracturing** (Fig. 11.2). Fluid is introduced at a rate sufficient to raise the downhole pressure above the fracture pressure of the formation rock. The stress induced by the pressure creates fissures and interconnected cracks that increase the permeability of the formation and enable greater flow rates of gas into the well. The fractures are held open by injecting small glass beads into the formation.

There is some small risk to overlying aquifers. Groundwater may be contaminated by extraction of shale gas, both from the constituents of shale gas itself, from the formulation and deep injection of water containing a cocktail of additives used for hydraulic fracturing, and from flowback water released during gas extraction, which may include saline formation water. Shale gas is predominantly methane of thermogenic origin with low percentages of ethane and propane. Documented instances of groundwater contamination from the United States all relate to the leakage of methane into groundwater.

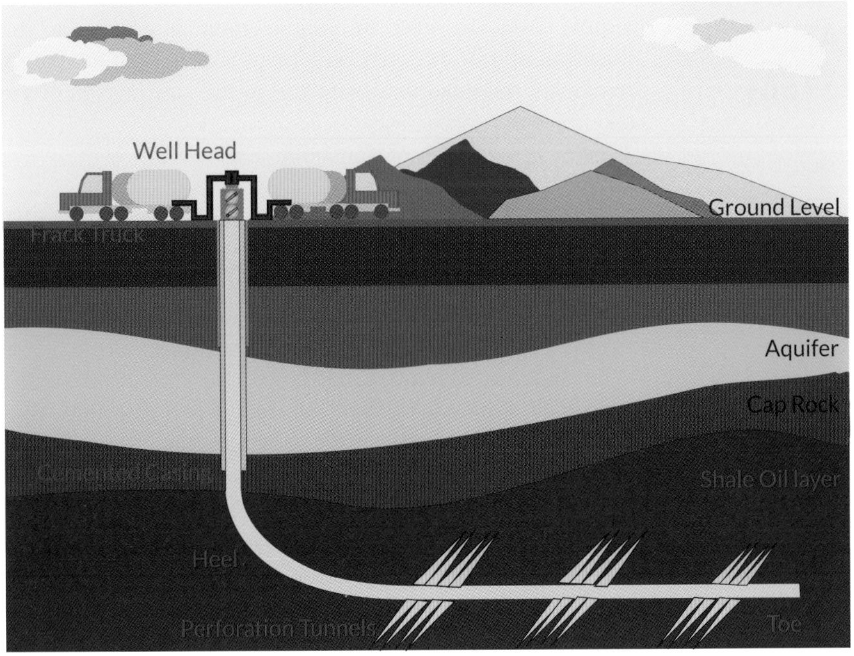

93

Figure 11.2 Schematic illustration of the oil shale facking procedure. Image credit: Shutterstock/BalLi8Tic.

The process of recovering the gas is very water-intensive and a large renewable resource is needed. The problem arises with the disposal of the return water, the poor quality of which, certainly in the United States, generally requires disposal by deep well injection. Nevertheless, the main risk to groundwater occurs where deep fracking takes place beneath a major supply aquifer to which released gas could potentially migrate.

Groundwater source heat pumps

Rocks, minerals and groundwater have a huge capacity to store heat. They have a near-constant temperature throughout the year. The rocks cool very slowly and they are generally warmer than the air in winter. Conversely, in summer rocks are generally cooler than the air. It is thus possible to extract some of this stored solar heat (and a component of genuine geothermal heat) via boreholes during the winter (Fig. 11.3). This heat energy may be tapped either by:

- Pumping groundwater from a borehole and extracting heat from it via a heat pump. This method is best suited to permeable rocks and wells with a high yield.

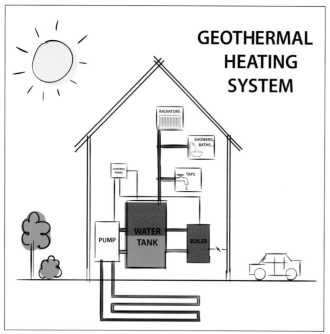

Figure 11.3 Schematic of the ground source heat pump concept. Image credit: Shutterstock/JeleR.

- Circulating a fluid through a closed pipe system down the borehole. The fluid is warmed to the temperature of the rocks and, on its return to the surface, may be sent through a heat pump.

A heat pump needs a small amount of electricity to run, and functions like a refrigerator. It takes heat from a low-temperature medium (e.g. British groundwater at c.10–12°C) and transfers it to a high-temperature space-heating medium at, say, 25°C. The electricity is used to 'push' the heat 'up' the temperature gradient. Heat pumps may be used to warm domestic properties, but are most effectively utilized for larger public buildings, or residential blocks. The energy gained far exceeds the energy used.

Transboundary Aquifers

The Transboundary Aquifer (TBA) is a groundwater unit shared by two or more nations or managing states within a nation, for which equitable resource apportionment is desired. One of the often reported examples is the management of the West Bank Mountain Aquifer, which is recharged in Palestine but flows as a confined aquifer to spring discharges in neighbouring Israel. Apportionment is controlled by Israel while Palestine, which Israel considers as an Occupied Territory, believes its resource allocation is inadequate. In most cases, however, the management of a TBA and the allocation of resources between neighbouring political units are carried out collaboratively and to the satisfaction of the stakeholders.

There are six types of transboundary aquifer:

- An unconfined aquifer that is linked hydraulically with a river, both of which flow along an international border (i.e. the river forms the border between two states).
- An unconfined aquifer intersected by an international border and linked hydraulically with a river that is also intersected by the same international border.
- An unconfined aquifer that flows across an international border and that is hydraulically linked to a river that flows completely within the territory of one state.
- An unconfined aquifer that is completely within the territory of one state but is linked hydraulically to a river flowing across an international border.
- A confined aquifer, unconnected hydraulically with any surface body of water, with a zone of recharge (possibly in an unconfined portion of the aquifer) that traverses an

international boundary or that is located completely in another state.

- A transboundary aquifer unrelated to any surface body of water and devoid of any recharge.

An example of cross-boundary derogation that illustrates the need to manage such aquifers is given by two neighbour states sharing an unconfined aquifer. One state is low-lying and coastal; the inland state is at a higher elevation. The inland state develops a citrus industry and plants orange groves which it irrigates with groundwater from the shared unconfined aquifer. The coastal state traditionally relies on date palms for its main crop. As the inland orange groves mature, depletion of the aquifer starts to occur so that the head of water at the coast is reduced and saline intrusion starts to occur beneath the date palms. The increased salinity of the water kills the date palms, destroying the main economic income of the coastal country; the inland country has an enhanced income at the expense of the coastal country. Another example is deep groundwater flow passing from state A to state B in a confined aquifer while shallow groundwater flows in the opposite direction

in the upper unconfined aquifer. TBAs are by no means straightforward.

Understanding a transboundary aquifer is underpinned by assessment of the overall surface and groundwater system. Classification and zoning of the respective aquifers is an essential prerequisite to prioritize management need. Standardized data collection, comparison and harmonization across borders are proving to be a key challenge. Classification of TBAs provides stakeholders with information necessary for decision-making and allows them to focus on those TBAs where co-operation and joint international management would help promote equitable division of the resource. TBAs can be classified as those resources with the potential to cause tension between neighbouring states, i.e. politically sensitive or troublesome, and those unlikely to become problematic even in the future. Stakeholders need to be armed with this classification in order to differentiate between TBAs that are likely to require careful management and those not currently in need of intervention.

Water wars

The links between water scarcity and public unrest are close (Fig. 11.4). In most arid and semi-arid environments, groundwater is the only viable resource capable of sustaining rural and urban populations. Periodic drought forces people off

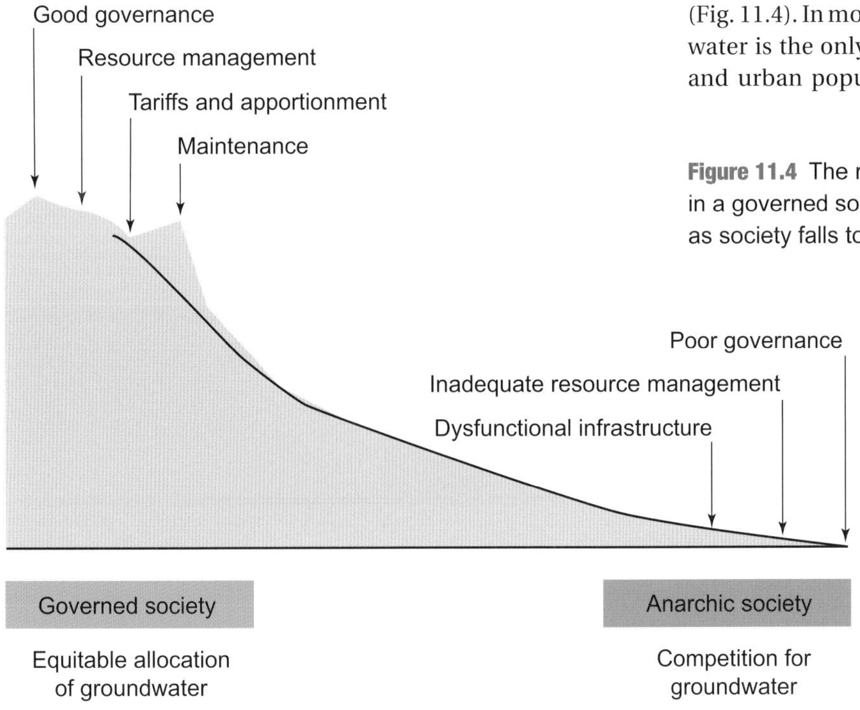

Figure 11.4 The relationship between groundwater availability in a governed society with the declining resource availability as society falls towards anarchy.

Good governance
Resource management
Tariffs and apportionment
Maintenance

Poor governance
Inadequate resource management
Dysfunctional infrastructure

Governed society

Anarchic society

Equitable allocation
of groundwater

Competition for
groundwater

the land; they then take refuge in the cities, taxing already stressed resources. Mining of groundwater places cities at risk; aquifers supplying both Yemen's capital city Sana'a and its major regional centre of Taiz are expected to run dry in the foreseeable future. In Somalia, al-Shabaab canvassed disfavour by mishandling the 2011 drought, which was partly self-inflicted through excessive tree-felling and a failure to repair broken infrastructure. Water scarcity was also a catalyst of the unrest in Syria, and has long been a critical factor in the tension between Israel and the Palestinians. Small island states and communities that live over transboundary aquifers are also vulnerable unless early technical investigation and collaborative interstate management are provided.

It is critical that the West helps all these afflicted countries. This is particularly so in places such as Yemen, Somalia and Syria, where the current instability stems partly from water scarcity issues, and where supply failure causes friction between competing tribal factions, leading to political instability. It is imperative that the focus is switched from fighting the symptoms of unrest, such as insurgent terrorist groups, to concentrate instead on the underlying causes such as poor governance and water resource scarcity (Fig. 11.5). Given the continuing unrest in those countries affected by the so-called 'Arab Spring', the link between water scarcity, invariably groundwater scarcity, and anarchy has become inarguable. The way to mitigate the ongoing threat to security must include technical advice on groundwater management and, where necessary, on preparing for the impacts of projected resource failure.

Figure 11.5 The ideal aid gift to a water scarce community: The Maharaja's Well at Stoke Row, Oxfordshire, dug with funds provided by the Maharaja of Benares to his retiring estate manager in 1871, complete with acknowledgment of the benefactor, a well-keeper's house and an orchard to fund maintenance work.

12   Hydrogeology

What does a hydrogeologist do?

The work of a hydrogeologist is applied to a variety of practical purposes. These include:

- Designing and testing water wells for drinking water supply, irrigation schemes and other purposes;
- Trying to discover how much water is available to sustain water supplies so that abstraction does not adversely affect the environment – for example, by depleting natural baseflows to rivers and important wetland ecosystems, or over-pumping and permanently damaging the aquifer;
- Investigating the quality of groundwater to ensure that it is fit for its intended use;
- Designing schemes to safeguard groundwater from pollution and clean it up when it becomes polluted;
- Designing dewatering schemes for construction and dealing with groundwater problems associated with mining;
- Helping to harness thermal energy through groundwater-based heat pumps;
- Advising on placement of hazardous waste underground and on innovative oil and gas exploitation.

Hydrogeologists are involved in attempting to solve some of the big questions facing the world today. These include sustainable water supply, food security and energy production, environmental protection and climate change. They work closely with a wide range of people, from individual farmers and well owners to the scientific and engineering community, as well as with agronomists, sociologists, economists, policy makers, regulators and planners. A major role is advising managers and policy makers across a wide remit of groundwater-related issues.

The subject has become so wide-ranging that hydrogeologists tend to work within a subset of the overall science. There are, for example, hydrochemists, isotope chemists, groundwater modellers, groundwater managers and regulators, legal advisors and many other specialists. Work invariably leads to the field, where basic observations and data gathering underpin all other activities. The fieldwork may be local, but more usually takes the hydrogeologist away from home for extensive periods, and sometimes includes work overseas. A large part of this activity is interactive dialogue with other workers, advising, sharing data and hypotheses, and exchanging experiences.

So how do I become a hydrogeologist? There are a number of routes to this goal, all of which involve a first degree in geology, environmental science, mathematics or related subject. A number of centres offer Masters degrees in hydrogeology, and this is the normal entry route to a career as a hydrogeologist. Initial job-openings include opportunities in the consulting sphere, perhaps focusing on contaminated land, dewatering or modelling support. Other opportunities arise with the regulators, with government agencies such as geological surveys and with local authorities. Perhaps the greatest honour bestowed on a hydrogeologist is the privilege of working overseas in a water-scarce area, helping to deliver fresh potable groundwater to poor communities (Fig. 12.1).

Figure 12.1 A satisfied community with a reliable water source in Malawi. Image credit: Jeff Davies.

The satisfaction of helping to improve the wellbeing of such a community and strengthening its ability to withstand periods of drought is immense.

How did the science begin?

Today hydrogeology is an important sub-set of the science of geology. It includes a broad field of interests that collectively analyse and describe the occurrence and distribution of underground water. In 1877 Joseph Lucas originally defined the science as:

> Hydrogeology … takes up the history of rainwater from the time that it leaves the domain of the meteorologist, and investigates the conditions under which it exists in passing through the various rocks which it percolates after leaving the surface.

The science of hydrogeology is not isolated from other sub-sets of geology. It relies on geophysics to locate favourable locations for drilling water boreholes, and after drilling it uses down-hole geophysics to identify productive formations. The hydrogeologist also works alongside the engineering geologist, both keen to know and understand the physical properties of the rock formations, permeable and weakly permeable. The sedimentologist helps predict facies changes in order to locate areas of high or low permeability; the mineralogist identifies likely chemical interactions with groundwater, both acid and alkaline; the structural geologist assists with predicting fracture patterns. Other sciences that are part of the hydrogeologist's toolkit include mathematics, physics and chemistry, but also surface water hydrology, climatology, sociology and economics.

The science of hydrogeology evolved in England over a period of about 400 years, starting in the seventeenth century with an increasing interest in mineral water springs and their occurrence. This interest questioned the theory that spring waters were seawater that had somehow been distilled on its pathway through the ground. Thereafter, scientific enquiry became the main driver, with experiments set up to test hypotheses that might begin to explain natural phenomena. However, by the eighteenth century more and more land had been put into agricultural production due to demographic expansion, and the era of land drainage commenced. This led to a broad understanding of the movement of water in the soil, the vadose zone, and in shallow unconfined aquifers.

The Industrial Revolution

The next driver was the Industrial Revolution and its increasing demand for water. Borehole drilling was a new focus, and success relied heavily on the burgeoning understanding of stratigraphy and the likely locations of permeable strata. Work was also taking place in France, where hydrogeologists had begun to identify the mechanics of basin-scale groundwater flow. In 1856 Henri D'Arcy, working in Dijon in France, was able to show the linear relationship between the flux and the hydraulic head gradient by using a wooden flow-through cell. D'Arcy concluded, 'it seems therefore that for a given sand, we can admit that the volume discharged is proportional to the head and inversely so, to the thickness of the layer crossed'.

The introduction of the steam-driven drilling rig was a great fillip to groundwater development, and the steam-driven beam pump became an essential component of water supply systems. A number of townships in Europe turned to borehole water for supply as their original spring and surface water sources became inadequate for the influx of people that moved off the land into the new industrial and urban centres. In England much of this early groundwater investigation and development was focused on the London Basin.

The twentieth century

In the early twentieth century little advance was made in Britain, while workers in Continental Europe and North America pressed ahead and made considerable progress. Agricultural demand in the United States was boosted by the introduction of the deep electric turbine pump in 1907, and understanding of confined and leaky confined aquifers led to O.E. Meinzer's seminal report on the compressibility of confined aquifers, published in 1928. In 1935 Theis published his work on unsteady state flow to a well. Not only were the Americans getting to grips with applied groundwater studies: their work was also becoming increasingly data-rich and was supported by numerical analysis.

Back in England it took a severe drought in 1930 to waken government to the need for hydrogeological research. A groundwater team was assembled within the Geological Survey, but effort was diverted to emergency water supply work in the Second World War. This led to a series of wartime pamphlets which, for the first time, catalogued the main aquifers of mainland Britain and the sources available from these aquifers. After the war, work in Britain was initially focused more on resource assessment work, but after 1963 the focus changed to groundwater resource management. Then in 1974 there began an intense effort to catalogue groundwater quality and to identify sources that were at risk from pollution. From 1990 onwards, the main effort has been integrated environmental management, with hydrogeology centre field in a variety of multi-disciplinary programmes.

Many of the current techniques for analysing groundwater flow were developed in the United States by people such as Hantush, Jacob and Cooper. Digital groundwater flow models were introduced with the advent of the mainframe computer in the 1960s, with work by Konikow and Bredehoeft leading to the publication of the MODFLOW code, the most widely used analytical groundwater flow code, which even now can be exchanged between international parties without either speaking a word of the other's language.

Hydrogeology has been a taught subject in universities since 1970. In England the two centres of hydrogeology were at Birmingham and University College, London. Taught at Masters level, the courses attracted numerous overseas students who took home with them a detailed knowledge of the science with which to develop hydrogeology all over the world. Expatriate hydrogeologists from Europe and North America, working in Africa, South and Central America, Asia and the Middle East were able to demonstrate applied hydrogeology. Many who did this say that this was how they learnt their trade, but as well as developing their own skills they were able to convince foreign governments of the importance of hydrogeology – albeit a fact that many governments now seem to have forgotten.

Hydrogeology today – an imperfect science

Hydrogeology is fast moving into a new, interdisciplinary phase where it is increasingly applied to projects and programmes designed to benefit people directly. Much of the current work is targeted at social wellbeing and wealth creation as well as regional and national economic gain and stability; the main goal of the hydrogeologist is sustainable exploitation of the resource. The two great challenges now are keeping supply in step with demand and anticipating the vagaries of future climate scenarios.

Digital data sets are increasingly available for public use. In the United Kingdom, for example, datasets are available for effective rainfall, land-use and soil type, river flow, morphology and several other associated fields. As monitoring technology improves with miniaturization of sensors, data storage and remote downloading, costs are reduced. These factors have combined to increase the number of monitoring stations supplementing the longer time series records that already exist. Remotely sensed data derived from instrumented satellites are increasingly being used to enhance inputs to groundwater models. These also provide a degree of international uniformity in data gathering, important for models that run across international boundaries. Ground-truthing is still essential, and field data gathering methods must underpin all other activities.

Analytical modelling

One consequence of the abundance of data is that modern hydrogeology has become dependent on the analytical model, and some practitioners have become blasé about the data they apply to the models and the historic water level and flow data they use to calibrate them. Groundwater modelling has traditionally addressed two areas: water resources problems and the fate and transport of solutes. However, hydrogeology cannot be an entirely quantitative science. Each groundwater model and each prediction is a hypothesis, and these predictions are rarely subjected to rigorous testing, other than by sensitivity analyses and 'ballpark' judgement as to how well the solution seems to fit the understanding of the aquifer. More often than not a complex numerical analysis of a problem is not needed, and a simple Darcian calculation will provide just as accurate an answer. Hydrogeology is mostly a descriptive science that attempts to be as quantitative as possible; this does not guarantee the accuracy of its predictions, but hydrogeologists currently strive to make their models more quantitative (rather than qualitative) in order to answer real management questions with better precision.

Modern hydrogeology continues to improve our understanding of basic principles and to solve practical problems. Quantitative science requires that hypotheses be tested, but hydrogeology is fraught by uncertainty, largely due to the heterogeneous nature of aquifers. This means that output from hydrogeological analysis is likely to be a best estimate, but because of the uncertainties, it is difficult to prove its accuracy.

A key problem is that the large-scale permeability of an aquifer should really be reconstructed from small-scale measurements rather than by **interpolation** between known values spread across a map. Interpolation to produce a generalized value still relies on point source measurements whose representativeness may also be questionable. Use of complex models of processes, or of models with complex parameter distributions and many ancillary details, is not always justifiable in order to answer practical hydrogeologic management and engineering questions. It is useful, however, for investigating hydrogeologic processes. The best use of numerical analysis in hydrogeology is to underpin qualitative descriptions in order to provide simple applied solutions to complex problems.

Given that current hydrogeological practice is still by no means perfect, we should look to the future to attempt to address these shortcomings.

Hydrogeology tomorrow

With the focus on human impacts and ecosystems, the natural sciences are becoming more interdisciplinary. A new realm of scientific investigation is emerging through the study of the connections and interactions between the atmosphere, hydrosphere, biosphere, cryosphere and solid Earth. Furthermore, scientists are starting to recognize the extent and magnitude of the anthropogenic impact on the entire Earth system. This requires a broadening and integration of the Earth sciences, with hydrogeology being just one important part of it.

The integrated environmental model

This holistic approach is the best way to address issues such as climate change, natural resource and energy security and the vulnerability of the environment. It will also highlight multi- and inter-disciplinary issues that require integrated understanding and analysis. We must consider the physical Earth system as a whole, bringing together not only climatic, ecological, hydrological, hydrogeological, and geological models, but also incorporating social and economic models. The concept of the integrated environmental model is one in which a holistic modelling platform allows interrogation between a variety of models. For example, a groundwater flow model could be run until it needs river flow data, the river flow model running until it needs climatic data to answer the groundwater flow models request, the solution returning to the river flow model, which then interrogates a demand model before it sends an answer back to the groundwater flow

model. Given the complexity of issues that are now starting to arise, this approach is likely to be the only sensible way to provide the necessary framework for future decision-making and management.

Understanding the geological setting of any hydrogeological province remains paramount to sensible interpretation of the hydrogeological regime. Using 3D visualization software as a platform to run analytical models enables 'what if' questions to be asked in the knowledge that the geological information is broadly correct, rather than, as so often happens, using an aquifer portrayed as a simple slab of porous medium with a defined top and bottom along with leakage characteristics and few, poorly defined vertical boundaries. Such estimation of reality is no longer acceptable, as it leads to serious errors and ultimately to the provision of misinformation to resource managers. Thus, a key step forward has already been made – that of sensibly and accurately capturing the geological framework in a 3D model as the foundation for the overall hydrogeological system.

The future is exciting. Hydrogeology is increasingly moving towards interdisciplinary studies involving climatological, environmental, social and other areas. These all require detailed interaction so that the social constraints and demands can be interrogated by the environmental needs, and climate change predictions can be loaded into any or all of the separate disciplines. This is now possible through a conjunction of sets of models, each able to interrogate the other and ask for data analyses to support the analytical programme of the next model. At the bottom of the pile of models sit the 3D data platforms, one for hydrogeology, perhaps one for hydrology and environment (ecology, habitat) one for climatology and one for sociology. Sounds complicated? Not really: work creating such modelling arrays is already well advanced. Nevertheless, human interaction must not be forgotten, and the output needs to be checked against simple logic and probability determination codes, with uncertainty quantified for each output.

Future sustainability

Groundwater sustainability will become an increasingly difficult area to address due to demography and the likely reduction in recharge due to climate change. Future research will depend on interdisciplinary science, predictive modelling, and interaction between science and society. The issues of heterogeneity need to be addressed. The distributed properties of porous-media and of secondary porosity or dual porosity flow systems so far elude unique resolution. However, there are two promising approaches for simulating heterogeneity in porous-media flow systems, one being facies modelling set within sequence stratigraphy, and the other being genetic, basin-scale sedimentation process modelling. Finally, the issue of uncertainty needs to be flushed into the open. It will no longer be adequate to say that the solution is such and such: in the future it will be necessary to report that the probable solution is X but that it ranges from V to Z.

The future of the science of hydrogeology is assured. There is much exciting work to be done within a new holistic analytical framework. The ultimate challenge is sustainability, and whether it can actually be achieved in many aquifer systems and groundwater basins around the world (Fig. 12.2). As one well-known hydrogeologist was recently heard to remark, 'the groundwater beneath the orange groves will never get contaminated by pesticides – the water table is going down year by year too fast for the chemicals to catch it up!'.

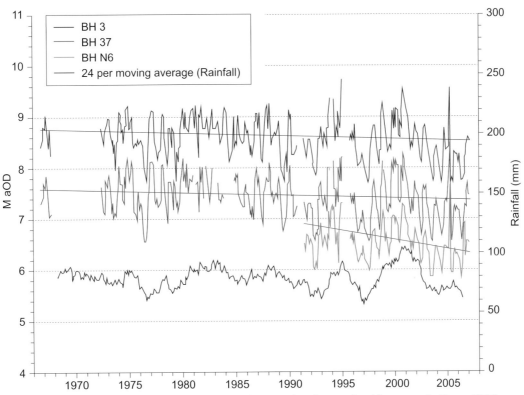

Figure 12.2 Might not seem a serious decline in water levels over the 40 year period from 1966 onwards but it is a real decline with real impacts on sustainability and society.

Glossary

A

acid mine drainage [56]: acid, metal-rich, discharge from coal and metal mines that have been dewatered allowing pyrite and other minerals to become soluble, and then later reflooded, so creating sulphuric acid and releasing iron and other metals in solution.

Actual Evapotranspiration (AE) [17]: the water flux which actually occurs. This is limited by the amount of moisture available in the soil. Estimates of Actual Evapotranspiration are derived from calculated values of Potential Evapotranspiration and current Soil Moisture Deficits (SMD).

aliphatic compound [67]: non-aromatic hydrocarbon.

annulus [40]: the space between two concentric cylinders, i.e. between the outside of a cased borehole and the drilled wall of the borehole.

antecedent [83]: preceding or prevailing conditions, can apply to rainfall, soil saturation or groundwater levels.

aqueous chemistry [1]: see **hydrochemistry**.

aquifer [5]: a consolidated or unconsolidated (saturated) geologic unit (material, stratum, or formation) or set of connected units that yields water to boreholes, wells or springs in economically usable amounts.

aquifer system [1]: intercalated permeable and poorly permeable materials that comprise two or more units separated by **aquitards** that impede vertical groundwater movement but do not affect the regional hydraulic continuity of the system.

aquitard [85]: a geologic material, stratum, or formation of low permeability (a confining unit) that can transmit water on a regional scale over geological time but does not permit significant water transfer across it locally.

artificial (engineered) recharge [12]: recharge caused by human design; injection of surplus water to a borehole or recharge lagoon. Where the same borehole is used to recover the water, the system is termed **artificial recharge and recovery**.

attenuation [67]: the process of reducing a quantity of solute or colloid, e.g. some form of pollution, in a groundwater system over time or space by both chemical and physical processes.

B

baseflow [3]: groundwater flow to a surface water body (lake, swamp, or stream); that portion of stream discharge that is derived from groundwater flow.

baseflow Index (BFI) [13]: the long-term percentage of surface water flow derived from groundwater in the catchment above a set datum, usually a gauging station.

biodegradation [67]: the transformation of a material to another material by organisms (commonly microbes).

borehole [9]: a cylindrical shaped hole drilled into the ground that penetrates the saturated zone and is of sufficient diameter to accommodate a pump or water level measuring device. See also **well**.

borrow pit [84]: a pit dig for the removal of material needed at another location.

bourne [81]: an ephemeral stream along an otherwise dry valley cut into Chalk. The bourne stream flows at times of prolonged wet weather in response to the water table rising up the valley.

C

capillary forces [2]: the action by which water is raised (or lowered) relative to the water surface because of interaction between the water molecules and the solids of the porous medium.

capillary fringe (or zone) [43]: the zone immediately above the water table where the medium is saturated or partially saturated by capillary rise from the phreatic zone.

capture zone [10]: the part of an aquifer that contributes water to a pumping well.

catchment [1]: the area of land drained by a single stream or river or, in the case of karst, drained by a single doline or group of dolines. Catchment divide and watershed are equivalent terms.

chloride mass balance [17]: an inexact means of determining recharge to a saturated or unsaturated zone by measuring groundwater chloride, or in the unsaturated zone by measuring chloride in soil water. Recharge is derived from total precipitation, the ratio of chloride concentration in rainfall and groundwater or in rainfall and the chloride concentration in the vadose zone.

circulation or residence time [26]: the time taken for a particle of water to travel the length of a flow-path from start to finish, i.e. the age of the water at its discharge point, or the time at discharge since it fell as precipitation and penetrated to the water table.

climate cycles [88]: the El Niño-Southern Oscillation, Pacific Decadal Oscillation, and Atlantic Multi-decadal Oscillation are the main climate cycles, and each has a periodic influence on global weather patterns.

conceptual model [35]: an initial schematic of the prevailing groundwater flow system in an aquifer unit, often combined with coarse estimate of the water balance.

cone of depression [51]: a curved conical-shaped water table or potentiometric surface that forms around a pumping well.

confined (or artesian) aquifer [6]: an aquifer that is immediately overlain by a low-permeability unit (confining layer). A confined aquifer does not have a water table.

conformal map [35]: in mathematics, a conformal map is a function that preserves orientation and angles locally. In the most common case, the function has a domain and a shape. Conformal maps preserve both angles and the shapes of infinitesimally small figures, but not necessarily their size or curvature.

conjunctive use [10]: the combined abstraction of surface water resources and groundwater, in a unified way, which optimizes the use of the resources.

connate water [5]: water that is in the pores at the time of sediment deposition.

constant head boundary [74]: a boundary condition, for example, in a computer model, along which the water level or head neither rises nor falls – the head is thus constant.

D

Darcy's Law [25]: the discharge of water (Q) through a unit area of porous medium is directly proportional to the hydraulic gradient (i) normal to that area (A). The constant of proportionality is the hydraulic conductivity (K). Thus Q = kiA.

data truthing, ground truthing [43]: using available data from boreholes and other sources to calibrate geophysical information.

derogate [52]: an action that reduces the quality and/or quantity of a groundwater source such as a borehole or spring.

determinand [62]: a hydrochemical or hydrobiological property of groundwater determined in the laboratory, inorganic, organic or biological.

diffusion equation [32]: a groundwater flow analogy to that used in heat transfer describing the flow of heat in a solid by conduction.

digital imagery [21]: a numerical representation, normally binary, of a two-dimensional image.

dilated fracture [22]: a fracture in which adjacent fracture walls, for the most part, are not in contact.

disperse pollution [56]: a pollution hazard that is dispersed across the land surface rather than a point source hazard.

DNAPL [67]: a dense non-aqueous phase liquid that is denser than water.

doline [89]: a round-shaped hole in chalk landscape caused by collapse of the ground over a void space created by dissolution; they are typically up to 10 m in diameter and 5 m deep.

drawdown [42]: the dynamic reduction in head from the initial static head caused by pumping from a borehole or well.

drinking water standards/guidelines [53]: prescribed maximum admissible concentrations of organic and inorganic substances present in domestic water supply. For example, nitrate as nitrogen is set at 11.3 mg l^{-1}.

dual porosity [26]: describes a media that has both intergranular and fracture porosity.

E

ecological (ecosystem) services [1]: ecological or ecosystem services are the varied benefits that are gained from properly functioning ecosystems in the natural environment. They include agro-ecosystems, forest ecosystems, grassland ecosystems and aquatic ecosystems. Collectively, these benefits are known as 'ecosystem services', and are often integral to the provisioning of clean and wholesome water resources.

effective porosity [23]: the interconnected porosity that contributes to groundwater flow. Often used synonymously with specific yield, although the two terms are not strictly the same.

effective rainfall [15]: the actual rainfall minus evaporation, less any surface water runoff.

effluent stream [4]: a surface stream or river receiving groundwater baseflow that is discharged into it.

evaporation [15]: the rate of water loss from a free water surface such as a reservoir, lake, pool, or saturated soil.

evapotranspiration [12]: the total water flux into the atmosphere, i.e. the sum of evaporation and transpiration (water flux through plant stomata).

F

field capacity [16]: the amount of water a soil can hold under natural conditions by capillarity and the suction of plant roots. If the water content is greater than the field capacity, then gravitational flow can occur. Also called field moisture capacity.

finite difference method [36]: a computed approximation of a continuous (groundwater flow) system as a grid of finite, commonly rectangular cells.

finite element method [36]: a computed approximation of a continuous (groundwater flow) system as a grid of finite multiple polygonal cells.

finite volume method [36]: similar to the finite difference method, but values are calculated at discrete places on a meshed geometry. Finite volume refers to the small volume surrounding each node point on a mesh.

flow-net [26]: a plan or map showing both equipotential lines and flow-lines (the path a molecule of water takes in its movement through a porous medium) in an aquifer. Used for evaluation of arcuate groundwater flow.

flux [36]: the rate of groundwater flow per unit area of porous or fractured media measured perpendicular to the direction of flow, i.e. inputs or outputs to the groundwater system.

formation constants [32]: a generic term for the hydraulic conductivity and storativity of an aquifer.

fracking [90]: a colloquial term for artificially fracturing deep oil-shale beds in order to release gas to collector wells.

fracture [1]: a sub-planar discontinuity in a rock formed by mechanical stresses. A fracture is visible to the naked eye and is open (i.e., not filled with secondary minerals).

fracture porosity (secondary porosity) [22]: the porosity of fractures and other non-intergranular voids.

freshwater lens [75]: a generic term to describe a lenticular-shaped freshwater body over sea water beneath a small, near-circular low-lying island.

G

Geographical Information System (GIS) [21]: GIS is a system that digitally captures, stores, manipulates, manages, and presents spatial or geographic data.

geophysical logs [40]: a suite of continuous geophysical measurements recorded down an unlined borehole column to strengthen the geological knowledge of the strata penetrated.

Ghyben–Herzberg theory [74]: the ratio of the elevation of the water table to the depth of the saline interface beneath sea level in a coastal or island aquifer is about 1: 40 based on the relative densities of the two fluids.

gravel pack [41]: a graded, granular, material placed in the annulus between a well screen and the formation (borehole wall) in unconsolidated sediments. If appropriate grain sizes are used the pack can enhance the connectivity between the borehole and the aquifer.

ground truthing [43]: see **data truthing**.

groundwater [1]: water beneath the land surface; it is more specifically defined as phreatic water, or water beneath the water table and unconfined water.

groundwater dome [79]: a mound of groundwater contained within an unconfined aquifer caused by recharge over the aquifer and its formation constants.

groundwater drought vulnerability maps [87]: groundwater resource potential, prevailing rainfall and rainfall distribution patterns, and population distribution are brought together to identify areas and communities that are most and least vulnerable to drought.

groundwater flood [81]: surface flooding caused when the water table rises to a level above the ground surface.

groundwater flow equation [32]: the mathematical relationship that describes the flow of groundwater through an aquifer.

groundwater mining [12]: a generic term to describe an aquifer that is being over-abstracted to cause long-term depletion of the overall groundwater storage.

groundwater potential [18]: the likely potential of a groundwater resource to produce set volumes of water for a given area of a prescribed quality. Sometimes called groundwater harvest.

groundwater vulnerability [18]: a term used to describe the vulnerability of a groundwater body at any location to a surface pollution hazard.

H

hard groundwater [22]: a generic term used to describe a water type rich in bicarbonate and calcium that is not suitable for boiler make-up water. It is typically expressed as the concentration of $CaHCO_3$ (mg/l).

head (h) [6]: fluid mechanical energy per unit weight of fluid, which correlates to the elevation that water will rise to in a well. Also hydraulic head; elevation head – head due to the energy that is the result of gravity (measured by the elevation of the water relative to some datum).

head boundary [74]: see **constant head boundary**.

heat source pumps [90]: a mechanism designed to abstract heat stored in groundwater for its heating and cooling properties.

Henry's Rule [75]: the highest elevation of the water table in a long thin island or sand bar is proportional to the width, and in an island to the diameter.

homogeneous [23]: a medium with the same properties throughout its aerial extent.

hydraulic conductivity (permeability) (k) [23]: the volume of fluid that flows through a unit area of porous medium for a unit hydraulic gradient normal to that area.

hydraulic fracturing [93]: artificial dilation of fractures by injecting fluid to raise the downhole pressure above the fracture pressure of the formation rock. The induced stress creates fissures and interconnected cracks.

hydrochemistry [2]: the chemical interrelationship between groundwater and the minerals that contain it, and defining chemical types of groundwater, from fresh through brackish to saline as well as the dominant classes such as sodium bicarbonate, calcium sulphate and sodium chloride types.

hydrograph [13]: a chart depicting water level (for example, measured in a borehole) against time.

hydrolysis [67]: the separation of water molecules into hydrogen and oxygen atoms using electricity.

hydrostatic pressure [84]: the pressure that is exerted by a fluid (e.g. by groundwater in a porous medium or a fracture) that is at equilibrium at a given point within the fluid, due to the force of gravity.

I, J, K

immiscible surface [74]: a surface between two fluids that do not readily mix together; the surface forms a diffusion zone in the case of seawater and fresh water in a coastal aquifer.

influent stream [14]: a surface stream or river discharging water as recharge into an underlying aquifer.

interflow [14]: water that infiltrates the land surface and flows into a stream but never reaches the local water table. Flow is mostly through the upper soil horizon and may be assisted by field drains.

interfluve [12]: see **watershed**.

intergranular porosity [23]: the porosity between the grains of a sediment or sedimentary rock.

interpolation [100]: constructing new data points within a field of known data points.

intrinsic permeability (k) [25]: the permeability of medium independent of the type of fluid present (also called the absolute permeability).

isotropic [23]: a medium in which the properties are the same in the vertical dimension as in the horizontal direction. **Anisotropic** media have different properties in the two planes.

juvenile water [5]: water released to the near surface from magmatic discharge and from hydrothermal fluids.

karstic flow [7]: karstic flow occurs in rocks such as limestones that have undergone significant dissolution by groundwater flow and attrition and are characterized by sinks into which streams and rivers disappear underground, risings in which underground flow emerges at surface, and cave systems through which rapid water flow takes place.

L, M

lithosphere [4]: the crust and the portion of the upper mantle that behaves elastically on timescales of thousands of years or greater.

LNAPL [67]: a light, non-aqueous phase liquid less dense than water.

major ion [58]: the anions HCO_3, Na, Ca, SO_4, and cations Cl, NO_3, Mg, K and Si. Respective concentrations in groundwater samples are presented graphically – see **Piper Diagram**.

managed aquifer recharge [53]: underground storage of surplus water for later use at times of water scarcity.

mass balance equation [32]: a statement of accounting, for a given control volume, aside from sources or sinks, e.g. recharge or abstraction, mass cannot be created or destroyed.

meteoric water [2]: water derived from the atmosphere as precipitation; a component of groundwater that derives from direct precipitation recharge.

Millennium Development Goals [57]: an objective to provide potable water supply and better sanitation coverage for the world's poor.

millequivalents (equivalents per million) [58]: concentration of an ion in water by dividing the concentration in mg l^{-1} by the equivalent weight (atomic weight divided by the valence).

mineralized/mineralization [1]: a term used to describe groundwater that has undergone some water-rock reaction to become mineralized to some degree. All groundwaters have some mineral content.

model [1]: a simplified description of a system or process that can be used as an aid in analysis or design.

N, O

nitrate vulnerable zone [52]: delineates areas of agricultural nitrate pollution in order to satisfy the requirement of the EU Nitrate Directive (91/676/EEC), where nitrate levels exceed 11.3 mg-N/l or were expected to in the future.

organic compound [61]: any compound containing carbon, e.g. pesticides, herbicides, hydrocarbons, plasticizers, and many types of solvent. Conversely, **inorganic** is a compound that does not contain carbon.

overdraft [50]: a generic term indicating that abstraction from an aquifer is exceeding the resource potential. Synonymous with **over-exploitation**.

P

palaeowater [29]: a generic term to describe a groundwater with a long circulation time.

pathogenic organisms [10]: a disease-producing organism/microbe.

percussion drilling [9]: a borehole drilling system that repeatedly raises and drops a heavy drill string and chisel-like bit onto the bottom of the hole. Debris is removed periodically with a bailer.

permeability [15]: the ease with which a porous medium can transmit water or other fluids. See also **hydraulic conductivity**.

phreatic zone [4]: see **zone of saturation**.

piezometer, piezometric surface [12/46]: a pressure-measuring device. This is typically an instrument that measures fluid pressure at a given point rather than integrating pressures over a borehole column. The piezometric surface or potentiometric surface is a surface of equal hydraulic heads or potentials.

Piper diagram [58]: a graphic representation of the major ion chemistry of groundwater samples. The cations and anions are shown by separate ternary plots. The apexes of the cation plot are calcium, magnesium and sodium plus potassium cations. The apexes of the anion plot are sulphate, chloride and carbonate plus hydrogen carbonate anions. The two ternary plots are then projected onto a diamond as a matrix transformation of a graph of the anions and cations.

piston flow model [60]: an idealized flow system through the unsaturated zone whereby the volume that drops out to the water table at the bottom equals the volume recharged at the top.

pollute [15]: a generic term for materials on and near the land surface that are harmful to man (for example, pathogens, nutrients, pesticides and pharmaceuticals) and which can percolate down into an aquifer.

pollution [10]: any aspect of water quality (physical, thermal, chemical, or biological) that interferes with an intended use by man or the surface environment on discharge.

porosity, pore spaces [22]: the volume of the voids divided by the total volume of porous medium.

potable [1]: drinkable. Potable waters can be consumed safely, as they are free from pathogens and other harmful components.

Potential (or Reference) Evapotranspiration (PE) [76]: the water flux under non-limiting soil water conditions.

pre-consolidation stress point [88]: dewatering of a plastic porous medium to a point where the pore pressure can no longer hold the fabric in place and the medium collapses, causing subsidence of the ground surface.

R

radiometric dating [28]: the application of isotope decay to age-dating of groundwaters

recharge [5]: the process by which water enters a groundwater system or, more precisely, enters the phreatic zone.

redox, redox potential [29]: the oxidation state of a solution.

regolith [24]: the weathered zone on the upper part of, for example, a crystalline basement type rock.

river augmentation [54]: a management system that uses groundwater to augment river flow at times of low surface water discharge.

river basin [1]: a large geographical region drained by a network of rivers that coalesce into a single discharge into the sea.

rotary drilling [43]: a borehole drilling system that uses a toothed hammer at the end of a drill stem that is rotated to penetrate the formation. Debris is removed in the upwelling drilling fluid that is pumped down the drill stem to cool the drilling bit. The **rotary table** is the circular platform on the drilling rig that imparts the rotational movement to the drill stem.

runoff [3]: water from precipitation, snowmelt, or irrigation running over the ground surface to flow into rivers, lakes, or reservoirs to become a component of stream flow.

S

safe yield [50]: a generic term indicating the amount of groundwater that can be removed from an aquifer before permanent depletion of the resource takes place.

salinization [71]: degradation of the soil and subsurface by the accumulation of salts – often caused by irrigation with mineralized groundwater.

saturation [67]: is reached when SMD= −10mm, i.e. a water surplus of 10 mm. Positive SMD is below field capacity and rain can infiltrate to the capacity of the SMD amount. In a saturated soil, all of the available soil pores are full of water, but water will drain out of large pores under the force of gravity.

sink hole [88]: a sudden collapse of the ground where overburden has collapsed over a void space created by dissolution of carbonate rock.

SMOW [61]: Standard Mean Ocean Water is a standard defining the isotopic composition of fresh water.

Soil Moisture Deficit (SMD) [16]: the amount of rain needed to bring the soil moisture content back to field capacity.

sorption [67]: the general process by which solutes, ions, and colloids become attached (sorbed) to solid matter in a porous medium.

source protection zone [52]: protects abstractions used for public water supply and food and drink production by limiting hazardous activities within prescribed zones.

specific capacity [48]: the discharge rate of a borehole divided by the drawdown at an assumed steady state, typically taken to describe the situation 24 hours after pumping has commenced.

specific electrical conductivity [59]: the ability of water to conduct electricity, which is a function of the ionic concentration measured in micromhos/cm.

specific yield (sy) [23]: the volume of water that a saturated porous medium can yield by gravity drainage per unit volume of the porous medium.

spring, seepage [3]: a concentrated or rapid discharge (or issue) of water from the earth, a small discharge to surface in an area of boggy ground. Either may be the source of a stream.

steady-state flow [32]: the condition in which properties in a flow system are not changing with time.

storage [5]: water contained within an aquifer or within a surface-water reservoir.

storativity, storage coefficient (S) [24/25]: the volume of water released per unit area of aquifer for a unit decline in head. In a confined aquifer, S is essentially the specific storage (Ss) times aquifer thickness; in an unconfined aquifer, S is essentially equal to the specific yield or the effective porosity.

submersible turbine pump [40]: a borehole pump in which the electric drive for the pump turbine is set underwater adjacent to the pump. Usually driven by a three-phase electricity supply.

T

test pumping [45]: one of a series of techniques to evaluate the hydraulic properties of an aquifer by observing how water levels change with space and time when water is pumped from the aquifer via a borehole.

Theis equation [47]: the equation for radial transient flow to a well in an idealized confined aquifer.

thermally induced groundwater flow [26]: groundwater flow in large basins caused by thermal advection.

trace element [59]: elements found in solution in water at extremely low concentrations.

transboundary aquifer [56]: an aquifer that straddles an international border or which can influence or be influenced by actions taken unilaterally within another country.

transient flow [32]: the condition in which properties of a flow system vary with time.

transmissivity (T) [25]: the discharge through a unit width of the entire saturated thickness of an aquifer for a unit hydraulic gradient normal to the unit width. Sometimes termed the coefficient of transmissibility.

transpiration [2]: see **evapotranspiration**.

turbidity [10]: water discoloured by suspended particulate matter.

turlough [81]: is a depression in limestone terrain (typically west of the River Shannon in Ireland) in which a rising causes the depression to flood with groundwater in wet weather and acts as a sink when the groundwater recedes in dry weather.

U, V

unconfined (or water-table) aquifer [6]: the upper surface of the aquifer is the water table. Water-table aquifers are directly overlain by an unsaturated zone or a surface water body.

unconfined aquifer [11]: an aquifer that has a water table and implies direct contact from the water table to the atmosphere (through the vadose zone).

unconsolidated sediments [1]: strata consisting of loosely bound material such as sand, gravel or cobble beds.

vadose zone [4]: see **zone of aeration**.

viscosity [25]: the internal friction of a fluid. It describes the resistance of a fluid to flow.

volatilization [67]: the process by which a liquid or solid goes into a gaseous phase. This is a major factor in the attenuation of organic liquids in shallow groundwater systems.

W

waste repository [90]: engineered containment of hazardous waste located at depth in weakly permeable strata with no significant apparent natural flow pathway back to the biosphere.

water balance [12]: the long-term equilibrium state of an aquifer by which the total amount of water arriving in an aquifer unit from rainfall recharge and other means of ingress equals the total discharge from the aquifer.

water cycle [1]: also known as the hydrological cycle, describes the continuous movement of water on, above and below the surface of the Earth. Water vapour is evaporated from the oceans and inland surface waters and much of this water falls as precipitation onto the land. Some is then returned to the atmosphere by evaporation, some runs off overland to streams and rivers and thence back to the sea, and the remainder percolates below the soil zone to the water table where it joins an aquifer, or groundwater body, eventually to discharge to surface and to a surface water system, thence also back to the sea.

water rights [50]: a means of empowering a person or company with the right to abstract surface and groundwater up to a specified rate of abstraction. This may be in the form of a traditional rule or through an authorized licence.

water scarcity [10]: a generic term to describe areas in which there is not always adequate water, surface water or groundwater, to sustain livelihoods and wellbeing. Such areas are also termed 'drought prone'.

watershed [1]: see **catchment**.

water table [4]: a surface at or near the top of the zone of saturation (phreatic zone) where the fluid pressure is equal to atmospheric pressure. In the field, the water table is defined by the level of water in wells that penetrate into the upper part of the zone of saturation.

well [5]: any artificial excavation or borehole constructed for the purposes of exploring for groundwater or for injection, monitoring or dewatering purposes. Synonymous with borehole is some parts of the world (e.g. United States) and used in preference to the term borehole throughout the oil industry.

well efficiency [46]: the effectiveness of the hydraulic connection between a borehole column and the formation penetrated. Turbulent inflow creates poor efficiency.

wellfield [7]: a group of boreholes or wells located in a constrained area, which collectively draw on an aquifer. The borehole pumping regimes are managed to optimize abstraction from the groundwater resource.

well points [84]: shallow and closely-spaced tubes jetted into soft ground and connected in line to a suction pump to intercept groundwater flowing towards an excavation.

well screen [40]: a length of slotted pipe used in place of plain casing in a borehole, and placed adjacent to the more productive horizons.

wetting threshold [14]: the point at which land in an arid or semi-arid climate starts to allow ingress of water that may contribute to groundwater recharge.

Y, Z

yield (from a borehole or well) [9]: the amount of water pumped from a borehole or well at approximately steady state, i.e. with the drawdown remaining constant.

zone of aeration [4]: the unsaturated zone above the water table that has both air and water in its pores.

zone of saturation [4]: the zone in which all the pores are filled with water.

Further reading

Investigating Groundwater (2019) by Ian Akworth, International Association of Hydrogeologists International Contributions to Hydrogeology Series, No 29, CRC Press, Leiden: ISBN 9781138542495

Water wells and boreholes (2017) by Bruce Misstear, David Banks and Lewis Clark, 2nd edition, Wiley Blackwell, Oxford: ISBN 9781118951705

Field Hydrogeology (2016) by Rick Brassington, 4th edition, Wiley Blackwell, Oxford: ISBN 9781118397367

Use the Internet: visit, for example, the International Association of Hydrogeologists https://iah.org/ and in particular the educational section on that site, https://iah.org/education; search for groundwater or hydrogeology in general and specifically on sites such as the British Geological Survey and the United States Geological Survey.